T0222855

The Road to Einstein's Relativity

Following in the Footsteps of the Giants

The Road to Einstein's Relativity

Following in the Footsteps of the Giants

David Lyth

CRC Press

Taylor & Francis Group

Boca Raton London New York

CRC Press is an imprint of the
Taylor & Francis Group, an **informa** business

CRC Press
Taylor & Francis Group
6000 Broken Sound Parkway NW, Suite 300
Boca Raton, FL 33487-2742

© 2019 by Taylor & Francis Group, LLC
CRC Press is an imprint of Taylor & Francis Group, an Informa business

No claim to original U.S. Government works

Printed on acid-free paper
Version Date: 20190108

International Standard Book Number-13: 978-0-367-00253-4 (Paperback)
International Standard Book Number-13: 978-0-367-02285-3 (Hardback)

Library of Congress Cataloging-in-Publication Data

Names: Lyth, D. H. (David Hilary), author.
Title: The road to Einstein's relativity / David Lyth.
Description: Boca Raton, Florida : CRC Press, 2019. | Includes
bibliographical references and index.
Identifiers: LCCN 2018046936| ISBN 9780367002534 (pbk) | ISBN 9780367022853
(hbk) | ISBN 9780429400445 (ebook)
Subjects: LCSH: Relativity (Physics)--Popular works. |
Physics--History--Popular works.
Classification: LCC QC173.57 .L95 2019 | DDC 530.11--dc23
LC record available at https://lccn.loc.gov/2018046936

Visit the Taylor & Francis Web site at
http://www.taylorandfrancis.com

and the CRC Press Web site at
http://www.crcpress.com

Contents

Preface

My father apparently told my wife-to-be that I was interested in nothing except satisfying my own boundless curiosity. I hope that was a bit of an exaggeration, but it's true that for as long as I can remember I've thought about why things are the way they are. That made physics a natural choice of subject for university.

After going on to do a PhD in theoretical elementary particle physics, I spent almost half a century writing articles and books for my fellow scientists. I wrote some of them by myself, but more often I collaborated with colleagues. For the first three decades, the books and articles were about elementary particle physics. Then, after taking a break, I moved on to cosmology.

The first book that I wrote for the general reader came out in 2016. It's called *The History of the Universe*, and it explains cosmology with no mathematics in the main text. I couldn't resist, though, a mathematical appendix and the whole book's about nothing except science. The second book, that you're looking at now, goes further down the road away from academia. There's no mathematics at all and most of the book isn't even about science. Instead it's about the life and times of scientists. I hope you find it interesting.

For comments on the final draft I would like to thank my wife Margaret, my brother Peter and my son John.

CHAPTER 1

Introduction

If I have seen further than others, it is by standing on the shoulders of giants.
(Isaac Newton, 1576)

Almost everybody knows that Einstein was a famous scientist, and most people know that he put forward a theory involving something called Relativity. Few, though, know anything about Relativity and its long history. In this book, I describe the life and times of some of the scientists who paved the way to Einstein's theory, and say something about the theory itself.

The basic idea of Relativity is simple. It's the idea that movement, in and of itself, cannot be detected; that nothing is affected simply by movement.

Put differently, the idea is that it makes sense to talk about movement only if the movement is relative to something else. In everyday life the 'something else' is, of course, the surface of the Earth—the ground. To say that something is moving is to say that it's moving across the ground. On the other hand, we do recognise that the Earth goes in an orbit around the Sun. In that case we're defining the Earth's motion relative to the Sun. Also in that case, we lose no sleep over the fact that we can't feel the earth's motion. These days we take the idea of Relativity for granted!

Attitudes were very different in 1632 when Galileo introduced the idea of Relativity. Then, there was a debate about whether the Earth really moved around the Sun or whether, instead, the age-old view was correct that the Earth was stationary with the Sun was moving around it. The suggestion that the Earth moved around the Sun wasn't particularly new, having being made by Copernicus in 1543. Even so, most people still believed that the Earth was stationary with the Sun and the stars moving around it. One argument that people used, was to say that if the Earth were really hurtling though space we would feel its motion. Galileo introduced the idea of Relativity to counter that argument.

Galileo invited us to imagine that we are in the cabin of a ship sailing smoothly across the water. Let us suppose, he says, that within the cabin are flies and butterflies, and a fish tank. The flies and butterflies go equally in all directions and so do the fish in the tank. He also invites us to suppose that there's a container with a small hole in its bottom, from which water is dripping. Each drop will hit the floor at the point directly below the hole. These examples could be multiplied, and in fact nothing within the cabin is affected by the motion of the cabin. We can find out that the ship is moving, only by looking out of the window. Galileo argued that, in the same way, we can't feel the motion of the Earth. We can find out that it's moving only by making astronomical observations.

The basic idea of relativity, then, is that nothing is affected by movement. Nowadays this basic idea is replaced by a more specific statement; that the *laws of physics* are not affected by the movement of the region within which they are applied. It was Einstein who came up with that statement. He called it the Principle of Relativity, and it was the starting point for his theory of Special Relativity.

'The laws of physics' sounds harmless enough, even a bit boring. Don't be fooled though. The laws of physics, as we now have them, are truly awesome in their scope. They seem to apply to *practically everything that can ever happen* and would cease to apply only in the most extreme hypothetical situations.

The journey to our present laws of physics may be said to have begun in 1687, when Isaac Newton laid down his three laws of motion. To this day, Newton's laws of motion are presented to students as a bedrock of physics. Newton used his laws to produce the first theory of gravity, whose most dramatic application is to the solar system.

When thinking about the solar system, we usually take the sun to be stationary. But Newton's laws and his theory of gravity would still apply if we took the sun to be moving, and the entire solar system with it. Newton's laws, and his theory of gravity, are therefore consistent with the Principle of Relativity.

So far so good for the Principle of Relativity. It seems to run into trouble though when we come to some laws of physics that appeared soon after Newton's death. These laws involve electricity and magnetism. An immediate problem comes with Coulomb's law, which specifies the electric force between any pair of *stationary* charged objects. Not stationary with respect to something else but simply stationary! More laws followed, all seemingly at odds with the Principle of

Relativity. It was Einstein's achievement to produce the theory of Special Relativity which, despite appearances to the contrary, makes the laws of electricity and magnetism consistent with the Principle of Relativity.

Special Relativity ignores gravity. After Einstein had formulated Special Relativity he thought a lot about the nature of gravity. He knew that there was something wrong with Newton's theory because it fails to describe correctly the orbit of Mercury. After years of effort, he at last came up in 1915 with a new theory of gravity which he called General Relativity.

General Relativity is directly relevant for our lives, because it's used by satellite navigation systems (satnavs) such as the American Global Positioning System (GPS). A satnav uses radio signals from three or more satellites to determine its position, and General Relativity is needed to interpret those signals. If Newton's theory were used instead of General Relativity, satnavs would give the wrong position causing absolute chaos!

In this book, my idea is to tell the story of the journey towards Einstein's theories of Special and General Relativity by focussing on key figures. I deal first with Archimedes (c. 287–c. 212 BC). He was the outstanding mathematician of the ancient world, and he enunciated Archimedes' Principle which is the first-ever law of physics. Then I move on to some of the astronomers that paved the way for Einstein. First is Copernicus (1473–1543) who (eventually) convinced people that the Earth and the other planets were moving around the Sun. Next is Kepler (1571–1630) whose three laws tell us exactly *how* the Earth and the other planets move around the Sun. Then there's Galileo (1564–1642) who we've already met. After that is the towering figure of Isaac Newton (1643–1727). Along with much else he gave us the inverse-square law of gravity, from which he he derived Kepler's three laws.

After all that astronomy I come to people who worked on electricity and magnetism. First up is Hans Oersted (1777–1851) who discovered that a loop of wire with a current flowing through it acts just like a magnet. Second up is Faraday (1791–1867) who found that if you push a magnet in and out of a closed coil of wire you generate a current in the coil. (That's how an electricity generator works.) Faraday also found that if you switch on an electric current through a coil, it will generate a burst of electric current in a nearby closed coil. (That's how an electricity transformer works.)

Next is André-Marie Ampère (1775–1836) after whom the amp is named. He found, among other things, the force between pieces of wire that each carry an electric current. (That's the theory behind an electric motor.) Then we come to James Clerk Maxwell (1831–1879) who drew together all these discoveries, finding in the process that a beam of light consists simply of electric and magnetic fields. Finally, we come to Einstein himself, who was born in 1879 and died in 1955.

Every one of these people made earth-shaking discoveries, and some of them had very eventful lives. I hope you'll enjoy reading about them.

CHAPTER 2

Archimedes

Modern mathematics was born with Archimedes, and died with him for all of two thousand years. It came to life again only with Descartes and Newton. (Eric Temple Bell, in The Development of Mathematics, *1940.)*

2.1   BIOGRAPHY

'Give me a place to stand, and with a lever I will move the whole Earth'. Not the saying of a modest man but Archimedes had no reason to be modest. His achievements in mathematics were unparalleled until the advent of Renaissance mathematics in the fifteenth century, and he remains one of the greatest mathematicians of all time. He can also lay claim to being the first-ever physicist, through his treatise *On Floating Bodies*. Galileo often described Archimedes as 'superhuman', and Leibnitz wrote that 'He who understands Archimedes and Apollonius will admire less the achievements of the foremost men of later times'. High praise from two more intellectual giants!

Archimedes was Greek, and was born c. 287 BC in the city-state of Syracuse. He died there c. 212 BC when he was about 75 years old. Syracuse is on the island that we now call Sicily. Hold on, I hear you say! How can Archimedes be called Greek if he never set foot in Greece? Well, confusing though it may be, one calls anybody in the ancient world Greek, if they spoke Greek and lived in areas that were culturally Greek. In addition to Greece, the culturally Greek areas included, in Archimedes' time, Sicily and parts of southern mainland Italy, some settlements around the Black Sea and a settlement near the mouth of the Rhone. The Romans called the culturally Greek areas Magna Graecia.

You might notice that this chapter doesn't display an image of its subject, as do the other chapters. That's because no known image of Archimedes exists. There are busts of other philosophers such as Plato, but none of Archimedes.

When Archimedes was born, the Punic Wars were in full swing. The Punic Wars were between the empires of Rome, and of Carthage which was near the present-day city of Tunis. Some of the rulers of Syracuse favoured one empire and some the other. This state of affairs ended, when Archimedes was still a child, with the seizure of power in 275 BC by King Hiero II who favoured Rome. There followed a fifty-year period of peace and prosperity, during which Syracuse became one of the most renowned cities of the Mediterranean. That was when Archimedes did the work which made him famous.

As Archimedes lived so long ago we don't know an awful lot about him. We know that his father Phidias was some sort of astronomer, but we don't know anything for sure about Archimedes' early life. People think he might have studied in Alexandria, on the coast of what is now Egypt. That's because his contemporaries Conon of Samos and Eratosthenes of Cyrene were certainly there, and Archimedes referred to Conon as his friend while two of his works have introductions addressed to Eratosthenes.

Alexandria had been founded by Alexander the Great, and it would have been quite an experience to live there because it was the second most important city of the Mediterranean. It was eclipsed only by Rome in terms of size and wealth.

We know that Archimedes died during the Second Punic War, when Roman forces under General Marcus Claudius Marcellus captured the city of Syracuse after a two-year-long siege. We also know that he was killed by a Roman soldier, but we don't know exactly what happened. Here's what happened according to one account. When the city was captured, Archimedes was contemplating a mathematical diagram. A Roman soldier ordered him to come and meet General Marcellus, but he refused saying that he had to finish working on the problem. The soldier was enraged by this and killed Archimedes with his sword. Another account is more mundane. According to that account it was armed robbery; Archimedes was carrying mathematical instruments which the soldier wanted to steal. However it was done, General Marcellus was angered by the death of Archimedes, because he considered him a valuable scientific asset and had ordered that he not be harmed.

The first of those accounts of Archimedes' death suggests that the act of thinking made Archimedes almost oblivious to the outside world. That seems to be confirmed by the Roman author Plutarch, who, in about AD 100, wrote this in his *Parallel Lives*.

> While the servants were anointing him with oils, with his fingers
> he drew lines upon his naked body. So far was he taken from

himself and brought into ecstacy or trance, with the delight he had in the study of geometry.

There's a tantalising story about the tomb of Archimedes. In 75 BC, more than a century after the death of Archimedes, the Roman orator Cicero was serving as quaestor in Sicily. (A quaestor was the person in charge of the public finances.) He had heard stories about the tomb, but the locals weren't able to tell him where the tomb was. He eventually found it near the Agrigentine gate in Syracuse, in a neglected condition and overgrown with bushes. Cicero had the tomb cleaned up, and on it he saw a carving depicting a sphere inside a cylinder which illustrated one of Archimedes's discoveries. He also read some verses that appeared on the tomb. Unfortunately, excavations near the Agrigentine gate haven't yet found the tomb though they have found other tombs.

Given his fame nowadays, it's not surprising that Archimedes has been commemorated. A crater on the Moon is named Archimedes and so is a lunar mountain range. Also, the Fields Medal for outstanding achievement in mathematics carries a portrait of Archimedes, along with a carving of a sphere inside a cylinder. The inscription around the head of Archimedes is a Latin translation of something attributed to him which reads 'Transire suum pectus mundoque potiri' (Rise above oneself and grasp the world).

2.2 INVENTIONS OF ARCHIMEDES

Archimedes was responsible for quite a few inventions. One of them appears in the following story. Apparently, a crown had been made for King Hiero II so that he could give it to a temple. King Hiero had supplied the pure gold to be used, and Archimedes was asked to determine whether some silver had been substituted for gold. He knew the weight of the crown and he knew the volume of a piece of gold with that weight. What he needed to do, therefore, was to determine the volume of the crown. If the crown had some silver mixed with the gold, its volume would be bigger than expected because silver is lighter than gold.

Archimedes is supposed to have been getting into his bath when he realised how he could determine the volume of the crown. When getting into the bath, he noticed that the level of the water in the tub rose as he got in, and he realized that this effect could be used to determine the volume of the crown. All he had to do, was to immerse the crown in a tub brim-full of water, and measure the volume of the water that flowed out of the tub. That's when he's supposed (on the basis of absolutely no evidence) to have been so excited by the discovery that he jumped out of the bath and ran through the streets naked, crying 'Eureka!' meaning 'I have found it!'.

Well anyway, the test was duly conducted. It showed that silver had indeed been mixed with the gold, which would have been extremely bad news for the person who supplied of the crown.

That story was written down by the Plutarch in his *Parallel Lives*. Although it sounds plausible, some scholars doubt it. For one thing, it doesn't appear in the known works of Archimedes. For another, the volume of the water flowing out of the tub would have to be measured very accurately. So it's been suggested that Archimedes might instead have used Archimedes' Principle, which is part of every child's scientific education. This is the statement that an object immersed in a fluid experiences an upward force equal to the weight of the fluid it displaces. Here's how one might have used Archimedes' Principle to decide if the crown was pure gold. Well at least, how one might if one had access to lots of gold.

First one would balance the crown on a pair of scales with gold weights in the other pan. Then the apparatus would be immersed in water to see if the scales remained in balance. If the crown were pure gold its volume would be the same as that of the reference sample. As a result, it would experience the same upward force as the reference sample and the scales would remain in balance. But if the crown were not pure gold its volume would be different from that of the reference sample. In that case, it would experience a different upward force and the two would not remain in balance. Galileo considered it 'probable that this method is the same that Archimedes followed, since, besides being very accurate, it is based on demonstrations found by Archimedes himself'.

The story of the crown is quite entertaining. The rest of Archimedes' inventions don't come with much of a story but I'll list them anyway. At one point, King Hiero II commissioned Archimedes to design a huge ship. The ship was called the *Syracusia,* and it could be used as a cargo ship, a naval warship or simply for luxury travel. The *Syracusia* is said to have been the largest ship built in classical antiquity. It could carry 600 people and it contained a decorated garden, a gymnasium and a temple dedicated to the goddess Aphrodite. When the ship was built, Archimedes designed a long screw to launch it.

That screw's not to be confused with the device called Archimedes' screw, which he had caused to be installed to remove bilge water from the ship. An Archimedes screw consists of a revolving screw-shaped blade inside a cylinder. When the blade is turned it transfers water along the cylinder. Archimedes' screw has also been used to transfer water from a low-lying body of water into irrigation canals, and is still used today for pumping liquids and granulated solids such as coal and grain. Archimedes' screw may have been an improvement on a device that we know was used to irrigate the Hanging Gardens of Babylon.

The propeller of a ship is an example of Archimedes' screw, which is why it's often called a screw. In this case the screw doesn't move the water much, but instead it mostly moves the ship through the water. The world's first seagoing steamship with a propeller was launched in 1839 and named the SS *Archimedes*.

Another device, known as the claw of Archimedes, was a weapon that he is supposed to have designed in order to defend the city of Syracuse. It consisted of an arm from which a large metal grappling hook was suspended. When the claw was dropped onto an attacking ship, the arm would swing upwards lifting the ship out of the water and possibly sinking it. There have been modern experiments to test the feasibility of the claw, and in 2005 a television documentary entitled *Superweapons of the Ancient World* showed the building of a version of the claw and concluded that it was a workable device.

Continuing with the theme of war, Archimedes might have used mirrors to produce a heat ray which burned ships attacking Syracuse. The second century AD author Lucian, living in what is now Syria, wrote that during the Siege of Syracuse Archimedes destroyed enemy ships with fire. In the sixth century AD, Anthemius of Tralles mentions burning-glasses as Archimedes' weapon.

There has been a debate about the credibility of Archimedes' supposed weapon ever since the Renaissance. Modern researchers have attempted to recreate the effect using only means that would have been available to Archimedes. One test was carried out in 1973 by the Greek scientist Ioannis Sakkas. It took place at the Skaramagas naval base outside Athens. An array of 70 mirrors was used, each mirror having a copper coating and a size of around 1.5 by 1 m. The mirrors were placed so as to constitute, collectively, a parabolic mirror which focussed the sun's light into a tiny region. In this region a plywood mock-up of a

Roman warship was placed. When the mirrors were focused accurately, the ship burst into flames within a few seconds. The plywood ship had a coating of tar paint, which may have aided combustion, but a coating of tar would have been commonplace on ships at the time.

In October 2005, a group of students from the Massachusetts Institute of Technology carried out an experiment with 30 cm square mirror tiles, focused on a mock-up wooden ship at a range of around 30 m. Flames broke out on a patch of the ship, but only after the ship had remained stationary for around ten minutes. It was concluded that the device was perhaps a feasible weapon, but that a more likely use of the mirrors would have been blinding, dazzling or distracting the crew of the ship.

In the modern era, devices like the heat ray are in actual use. One is the heliostat, which is a moving plane mirror that directs the sun's light onto a window for an extended period. Another is the solar furnace, which strongly focusses sunlight and can be used for a variety of purposes.

Continuing with Archimedes on a more peaceful note, he built devices which displayed the motion in the sky of the Sun, the Moon, Mercury, Mars, Venus, Jupiter and Saturn. General Marcus Claudius is said to have taken two of the devices back to Rome after the capture of Syracuse c. 212 BC, He's said to have kept one of them, and donated the other to the Temple of Virtue in Rome.

Constructing such a device would have required a sophisticated knowledge of differential gearing. According to Pappus of Alexandria, Archimedes wrote a now-lost book *On Sphere Making*, explaining how to do it. An extremely sophisticated version of the device, called the Antikythera device, was discovered in 1902. It had been built in about 100 BC, and it employed technical knowledge that was subsequently lost, to be recovered only in the fourteenth century. To my mind that sheds a whole new light on Archimedes; he was a superb mathematician, a prolific inventor *and* a very skilled technician.

We're still not quite finished with Archimedes the inventor. He designed pulley systems to load and unload cargo from boats. He improved the power and accuracy of the catapult, for use against the Romans during the siege of Syracuse. He also constructed the first known odometer (a device for measuring distances). It consisted of a cart, with a gear mechanism that dropped a ball into a container after each mile. Such a device remained in use in Europe until about AD 200, but was then lost until the middle of the fifteenth century.

2.3 ARCHIMEDES' WRITINGS

Archimedes' inventions are impressive, but they're not why he's in this book. What has made Archimedes truly immortal are his writings. They consist of several mathematical works and a work on physics. I'll list all of them (or anyhow, all that are known) along with some comments.

On the Equilibrium of Planes explains the Law of the Lever. This law is easy to understand; it states that objects placed on a lever are in equilibrium, if their distances from the point of support of the lever are inversely proportional to their weights. Everyone who's ever used a see-saw knows that; if you put your kid on one end, you yourself had better sit a good way off the other end.

In this statement of the law of the lever, the 'distances' are actually distance from the point of support to the *centres of gravity* of the objects. Archimedes calculated the areas and centres of gravity of various flat shapes, including triangles and parallelograms. I don't think he calculated the centres of gravity of solid objects though, which one needs in real life.

In *On the Measurement of a Circle* Archimedes obtains an approximation for the number π, defined as the circumference of a circle divided by its diameter. He also shows that the area of a circle is π times the square of its radius. Those results are perhaps not surprising, but here's something that *is* surprising. Archimedes states that $\sqrt{3}$ is between $265/153$ (which to eight significant figures is 1.7320261) and $1351/780$ (which is 1.7320512). The actual value is 1.7320508, which means that Archimedes' result was accurate to about one part in a million. Archimedes didn't say how he got the result, which led the seventeenth century English mathematician John Wallis to remark that Archimedes was: 'as it were of set purpose to have covered up the traces of his investigation as if he had grudged posterity the secret of his method of inquiry while he wished to extort from them assent to his results'.

On the Sphere and Cylinder discusses the volumes of a sphere and of a cylinder, and their surface areas. It is shown that a sphere has $2/3$ of the volume of a cylinder that just fits outside it, and that area of the sphere's surface is $2/3$ of the area of the cylinder's surface including the cylinder's bases. A diagram of the sphere and cylinder was placed on the tomb of Archimedes, because he had asked for that to be done.

On Spirals defines what is now called the Archimedean spiral. To draw this spiral one draws a line on a cylinder with a pen that moves *around* the axis of the cylinder with constant speed, and which at the same time moves *along* the axis of the cylinder with constant speed.

On Conoids and Spheroids calculates the areas and volumes of some geometrical shapes.

On Floating Bodies studies solid objects immersed in a fluid. This work goes beyond Archimedes' other works, which are concerned exclusively with geometry, and it may be regarded as the earliest writing on what we now call physics. It's this work that contains Archimedes' Principle.

The Quadrature of the Parabola contains, among other things, a comparison of two regions. One of them is the region enclosed by a horizontal straight line and a U-shaped curve called a parabola. The other is the right-angled triangle, whose longest side is that same straight line, and which points downwards so that the corner opposite the longest side just touches the parabola. (I hope you got that. If you didn't and it bothers you, I'd say that pen and paper are called for.) It's shown that the area of the first region is 4/3 times the area of the second region.

The Cattle Problem is given in the following words (translated from the Greek, and somewhat simplified).

> The sun god had a herd of cattle consisting of bulls and cows, which were white, black, spotted or brown. Among the bulls, the number of white ones was equal to the number of the black minus the number of the brown, times 5/6; the number of the black was the number of the spotted minus the number of the brown, times 9/20; the number of the spotted the number of the white minus the number of brown, times 9/20. Among the cows, the number of white ones was 7/12 times the number of black cattle; the number of the black was 9/20 times the number of spotted cattle; the number of spotted, 11/30 times the number of brown

cattle; the number of the brown was 13/42 times the number of white cattle. What was the composition of the herd?

Archimedes didn't solve this problem. It was first solved, to a good approximation, by the German mathematician A. Amthor in 1880, and the total number of animals in the herd is about 8×10^{206544}. Some herd!

The Sand Reckoner estimates that the Universe could contain at most 8×10^{83} grains of sand. The number itself isn't of interest, because it was based on quite crazy ideas about the Universe. What is of great interest, is that Archimedes defined the number using a power of ten. It's interesting because the Greek number system was similar to the Roman number system. In it, letters are used to denote the numbers $1, 2, \cdots 9$, $10, 20, \cdots 90$, $100, 200, \cdots 900$ and 1000. Then the other numbers are denoted by putting letters together, as is familiar to us from the Roman system. For example, 1723 in the Roman system is $MDCXXIII$. Systems like that are of little use in defining the number 10^{83}, which is why Archimedes expressed it using a power of 10.

The Method of Mechanical Theorems describes two methods for finding out areas and volumes.

2.4 SOURCES

The earliest accounts of Archimedes were written well after his death, by historians of ancient Rome. The first was written by Polybius in c. 140 BC, in his *Universal History*. That was the source of accounts by Livy and Plutarch in the first century AD, and it sheds little light on Archimedes himself. Instead it focusses on the war machines that he's said to have built in order to defend Syracuse. Another account was given by Marcus Vitruvius in about 20 BC, in *Ten Books of Architecture*.

The statement that Archimedes might have been related to King Hiero II comes from *Parallel Lives* written by Plutarch in about AD 100. That work also contains the story of the golden crown and the two descriptions of Archimedes' death. It also contains the statement that Archimedes designed a pulley system for handling ships' cargos.

Archimedes' device for displaying the motion of heavenly bodies is mentioned by Cicero in his dialogue *De Republica* which portrays a fictional conversation taking place in 129 BC. Cicero also mentions predecessors of Archimedes' device, designed by Thales of Miletusin about 600 BC and by Eudoxus of Cnidus in about 350 BC. According to Cicero, Marcellus' copy of the device was demonstrated by Gaius Sulpicius Gallus to Lucius Furius Philus, who described it in these words (translated from the Latin)

> When Gallus moved the globe, it happened that the Moon fol-
> lowed the Sun by as many turns on that bronze contrivance as in
> the sky itself, from which also in the sky the Sun's globe became
> to have that same eclipse, and the Moon came then to that po-
> sition which was its shadow on the Earth, when the Sun was in
> line.

In the fourth century AD, Pappus of Alexandria stated that Archimedes had written a manuscript on the construction of these mechanisms entitled *On Sphere Making*. That manuscript has not been found. Pappus also attributed to Archimedes the remark that I cited at the beginning of this chapter. Archimedes' commission to design the huge ship is described by the Greek writer Athenaeus of Naucratis in about AD 200. He also made the statement that it could carry 600 people.

Now I come to Archimedes' own works. Archimedes wrote in Doric Greek, the dialect of ancient Syracuse. During his lifetime Archimedes made his findings known through correspondence with the mathematicians in Alexandria. In contrast with his inventions, his written works were little known in antiquity. The first comprehensive compilation was made in about AD 530 by Isidore of Miletus in Byzantine Constantinople (now Istanbul). Isodore was one of the two main designers of the Hagia Sophia cathedral in that city. A commentary on the works of Archimedes was written by Eutocius of Ascalon at about the same time. (Ascalon was at that time a Greek city-state. It is now Ashkelon in present-day Israel.) The relatively few copies of Archimedes' written work that survived through the Middle Ages were an influential source of ideas for scientists during the Renaissance.

Another source for Archimedes' writings, which was not available to scholars in the Middle Ages, was found in 1906 in the Archimedes palimpsest. A palimpsest is a document that has writing superimposed on older writing. In this case the older writing was a tenth century Byzantine Greek copy of works by Archimedes, and the more recent writing consisted of a Christian religious text written by thirteenth century monks. The original writing was recovered by shining onto the document a beam of light that was almost parallel to its surface, as well as by shining X-ray, ultraviolet and infrared light onto the document.

Except for the palimpsest, the works of Archimedes that we know about have survived because they were translated into Arabic by Thabit ibn Qurra in the ninth century, and into Latin by Gerard of Cremona in the twelfth century. During the Renaissance the *Editio Princeps* (*First Edition*) was published in Basel in 1544 by Johann Herwagen. It contained works of Archimedes in Greek, and in Latin translation. The palimpsest contains in addition *The Method of Mechanical Theorems* which was not previously known.

For further reading about the topics mentioned, see the following; biographies of Archimedes [1, 2, 3]. Archimedes' writings [4], Pappus of Alexandria [5], Plutarch's *Parallel Lives* [6, 7], Alexander the Great [8], Polybius' *Universal History* [9], Vitruvius' *Ten Books of Architecture* [10], Cicero's *Tusculan Disputations* [11], the work by Athenaeus [12, 13], the Hanging Gardens of Babylon [14], the experiments on Archimedes' heat ray [15], Cicero's *de Republica* [16], history of Greek mathematics [17].

CHAPTER 3

Copernicus

The fool will upset the whole science of astronomy. (Martin Luther in Table Talk, *AD 1566.)*

3.1   INTRODUCTION

Copernicus is famous because he boldly declared his belief that the Earth is turning once every 24 hours, and is going around the Sun once a year. This statement, which I'll call the heliocentric theory, stood on its head the universal belief that the earth isn't moving. It led to what's often called the Copernican Revolution which was a complete alteration in our view of the Universe and of our place in it

Before coming to Copernicus I give some background. A lot happened in Europe, between the death of Archimedes in 212 BC and the birth of Copernicus in AD 1473. The Punic Wars ended in 146 BC, with the total victory of Rome and the destruction of Carthage. The Roman Empire, at its greatest extent, stretched from Britain to northern Iraq and from Algeria to Egypt. It spread to these regions many things, including Christianity which became the only permitted religion in AD 398. Soon after that the western part of the empire started to fall apart with the invasion of Germanic tribes, and it ended in AD 476.

The learning of the ancient Greeks was then largely lost in Europe. Much of it re-appeared in Spain after its invasion in 711 AD of the Arabs. They had picked it up in the countries bordering the Eastern end of the Mediterranean. As well as reviving the Greek learning the Arabs brought to Europe learning from other places. Most notably they brought the decimal system for whole numbers, which had first been developed by mathematicians of the Indian subcontinent around AD 500. These things provided the starting point for the intellectual explosion of the Renaissance which began in the fourteenth century.

An important religious development was the division of Christian Europe between Catholics and Protestants. (By specifying 'Christian' I mean to exclude the Balkans which were then ruled by the Ottoman

Turks. Most of the Balkan Christians were either Greek Orthodox or Russian Orthodox.) The Protestant movement was begun in 1517 by Martin Luther in Eisleben in northern Germany. By the end of the sixteenth century it had spread (adopting modern boundaries) to the whole of northern Germany, Great Britain, Scandinavia, the Netherlands, and parts of Switzerland and Poland. The Catholic areas were southern Germany, the rest of Switzerland and Poland, Ireland, France, Belgium, Austria, Spain, Portugal and Italy.

3.2   BIOGRAPHY

Nicolaus Copernicus

Nicolaus Copernicus was christened Mikolaj Kopernik, of which Nicolaus Copernicus is the Latinized form that we now use. He was born on 19 February 1473 in Thorn, which is modern-day Torun on the river Vistula. He died on 24 May 1543 in Frombork. Both towns were in Royal Prussia, a region bordering on the Baltic Sea that had been part of the Kingdom of Poland since AD 1466.

Nicolaus' father was a merchant and a politician. He had supported Poland against the Teutonic Knights, a force of soldier-monks that had in the thirteenth century converted the pagan Prussians. In 1454 he mediated negotiations between Poland's Cardinal Zbigniew Olesnicki and the Prussian cities for repayment of war loans.

Nicolaus' mother, Barbara Watzenrode, was the daughter of a wealthy Torun aristocrat and city councillor called Lucas Watzenrode the Elder. Through the Watzenrodes Copernicus was related to prominent families. Lucas and Katherine had three children; Lucas Watzenrode the Younger who would become Bishop of Ermland and Copernicus's patron, Barbara the astronomer's mother, and Christina.

Lucas Watzenrode the Elder had been a prominent opponent of the Teutonic Knights. In 1453 he was the delegate from Torun at the Grudziadz conference that planned the uprising against them. This had led to the Thirteen Years' War between the Teutonic Knights and Poland, during which a large fraction of the population died. During that war he had given financial support to the Prussian cities, only part of which he later re-claimed, and he had also fought in battles. He had died in 1462, and the Thirteen Years' War had ended in 1466 when the Teutonic Knights once and for all abandoned their claim to Royal Prussia.

Notwithstanding his Polish baptismal name, Copernicus' mother tongue seems to have been German which he spoke fluently as well as Latin. Later he also became fluent in Greek and he knew some Polish and Italian too. He was the youngest of four children. His brother Andrew became an Augustinian canon at Frombork. His sister Barbara became a Benedictine nun and ended up as the prioress of a convent in Chelmno. His sister Katharina married the businessman and Torun city councilor Barthel Gertner. She died leaving five children, whom Copernicus looked after for as long as he lived.

Copernicus' father died when he was ten. Lucas Watzenrode the Younger, whom I'll call simply Watzenrode, then took the boy under his wing and saw to his education and career. Watzenrode had fathered an illegitimate son but it was Copernicus that he considered to be his heir.

In 1491, when Copernicus was eighteen, it was time to go to a university. Young men from Torun generally went to Krakow, Leipzig or Prague University. Prague was being disfavoured at the time because of the Protestant movement started there by John Hus. For Copernicus, Krakow was a better choice than Prague for several reasons. His married elder sister was there. It was his father's old home town, giving him connections. Watzenrode had gone to university there. Copernicus therefore enrolled at Krakow University. From the records, it seems that his brother Andrew probably enrolled at the same time.

At Krakow University there were students from quite a few European countries. They conversed in Latin, which all educated people at the time knew, and it was there that Copernicus gave himself that Latinised name that we now use. He studied, logic, rhetoric, physical science and mathematical astronomy. While at Krakow Copernicus started to build up a collection of books on mathematics and astronomy. Some of them survive, and contain jottings which suggest that, even then, Copernicus was beginning to reject the received wisdom of astronomy. If so it was an early display of his creative thinking, because a university course at that time centered on the writings of ancient Greek and Roman authors. When there were formal debates on a subject the focus was on those writings rather than on facts.

Copernicus probably left Krakow University without a degree, as did most students in those days. He had made friends among the teachers and fellow students, and he kept up a correspondence with them.

After leaving the university he joined the court of his uncle Watzenrode who, in 1489, had been elevated to Prince-Bishop of Ermland. Ermland was an enclave within a territory to the east of Royal Prussia that was otherwise occupied by the Teutonic Knights.

In 1496 Watzenrode sent Copernicus to Bologna to study Canon Law. Canon Law is the legal code of the Catholic Church, and Bologna was the best place in Italy to study it. Italy was also the place where one could best absorb the culture of ancient Greece and Rome, and learn Greek from a native speaker.

Bologna University was at that time firmly in the hands of the students. A lecturer was answerable to the Rector who was the student's elected leader. He could be fined for all kinds of offences and if he married his honeymoon was limited to one day! Copernicus started his studies in the autumn of 1496. He was joined late in 1498 by his brother Andrew who also came to study Canon Law. They depended on their uncle for money, which they seem to have spent rather too freely. It was a great start in life for the two brothers, which they could not have known would end in triumph for one and tragedy for the other.

Instead of actually focussing on Canon Law, Copernicus became the disciple and assistant of the famous astronomer Domenico Maria Novara da Ferrara. He also developed new ideas after reading *Epitome in Almagestum Ptolemei* (*Epitome of Ptolemy's Almagest*) by George von Peuerbach and Johannes Regiomontanus. He verified its remarks concerning Ptolemy's theory of the Moon's motion, by conducting on 9 March 1497 at Bologna an observation of the passage of the Moon in front of the star Aldebaran.

Copernicus studied at Bologna for three and a half years. Then came the year 1500 which the Church was celebrating as a Jubilee Year. People from all over Europe went to Rome, and among them were Copernicus and his brother. They would have been among the 200,000 people who, on Easter Sunday afternoon, knelt to receive the blessing of the Pope. It was while they were in Rome that the Pope's son Cesare Borgia had the husband of his sister Lucrezia murdered. With things like that happening I don't find it surprising that Martin Luther started the Reformation a decade later.

I've now mentioned sons of both a bishop and a pope! Although the Catholic clergy were supposed to be celibate, it wasn't that unusual at the time for them to father children.

Copernicus stayed in Rome for most of 1500, giving some lectures on mathematics and astronomy. He also continued his astronomical observations.

In 1497 Copernicus had been made a canon of Frombork in Ermland. Frombork is on the Vistula Lagoon where the river Vistula meets the Baltic Sea. From then on, his expenses had been born by the cathedral instead of by his uncle. Copernicus had obtained leave of absence from Frombork. In 1499 his brother Andrew had also been made a canon, and he too had obtained leave of absence.

The brothers were installed in 1501 during a brief visit to Frombork, after which they received more leave of absence. Nicolaus' leave was to study medicine, 'since he may in future be a useful medical advisor to our Reverend Superior [Watzenrode]'. In late summer or autumn the brothers again crossed the Alps to Italy, accompanied now by their compatriot Canon Bernhard Sculteti who worked in the Vatican.

This time Copernicus studied at the University of Padua which was famous as a seat of medical learning. It's mentioned in Shakespeare's *Merchant of Venice* when a disguised Portia pretends to have come from Bellario, a 'learned doctor' of Padua. The change of university was allowed and even encouraged for students throughout the Middle Ages. Wherever they went they could pick up their studies from the point where they had left off. He received his degree in Canon Law in the autumn of 1501. He remained in Padua until the summer of 1503 except for a brief visit to Ferrara.

It was probably in Padua that Copernicus learned Greek. He read some of the classic Roman and Greek works, which are now widely available but then existed only in hard-to-find manuscripts. He also studied ancient Greek and Roman books on mathematics and astronomy. As well, he studied what passed for medicine in those days and acquired textbooks on the subject. Copernicus doesn't seem to have questioned the received (and wide of the mark) wisdom concerning medicine. In his lifetime he was thought of more as a physician than as an astronomer. As a medical student Copernicus would have studied astrology, because the casting of a patient's horoscope was thought to be important. Unlike most other Renaissance astronomers though, Copernicus doesn't seem to have been interested in astrology.

In the spring of 1503 Copernicus went to Ferrara where, on 31 May 1503, having passed the obligatory examinations, he was granted the degree of Doctor of Canon Law. One reason that he obtained his degree from Ferrara rather than from Padua, might have been the fact that a person receiving a degree was expected to throw a lavish party for his friends. Copernicus would have had many friends in Padua but few in Ferrara!

His leave of absence having ended, the 33-year-old Copernicus returned in 1505 or 1506 to Ermland. That was the end of his years of wandering. Instead of taking up his duties as a canon at Frombork though, Copernicus obtained further leave of absence to be the physician of Bishop Watzenrode in Lidzbark. In 1510 Copernicus became a canon in Lidzbark and lived in the bishop's castle. Canons were members of the cathedral Chapter which administered the cathedral, and they advised the bishop who was the Chapter's most senior member. There were altogether sixteen canons at Lidzbark, who had to conduct morning and evening prayers. As a canon he had a couple of servants and three horses. The canons owned about one third of Ermland, called the canonry, which provided their income.

Copernicus' time in Lidzbark ended in 1512 with the death of Lucas Watzenrode. He then took up his position as a canon in Frombork. His brother Andrew was already there, but Andrew soon contracted leprosy and died abroad within a few years.

Copernicus lived in Frombork for the remaining 40 years of his life. He was his uncle's secretary and physician from 1503 until at least 1510. He made numerous journeys with him, including journeys to sessions of the Royal Prussian diet (parliament). In April 1512 he took part in the election of Fabian of Lossainen as Prince-Bishop of Ermland. In early June 1512 Copernicus was given a house outside the defensive walls of the cathedral mount. In 1514 he purchased the northwestern tower within the walls. He would maintain both these residences to the end of his life.

In the same year Copernicus distributed a handwritten book to his friends that set out the heliocentric theory. Although he hadn't promoted himself he was by now well known as an astronomer. Later in the year, when the Catholic church was seeking to improve the calendar, one of the experts to whom the pope appealed was Copernicus.

Towards the end of 1516 Copernicus was appointed to manage two estates belonging to the Chapter. He therefore moved to Allenstein Castle about 50 miles inland from Frombork. From November 1519 to November 1520 Copernicus was back in Frombork. War had broken out between Poland and the Teutonic Knights which, as was usual at the time, basically meant that each side killed the other side's civilians.

As a side result of the war the town of Frombork was burned, but Copernicus was safe behind the Cathedral walls.

In November 1520 Copernicus resumed his duties in Allenstein. He witnessed a countryside that had been depopulated by war. All but two of the canons had fled and were haggling over money. The war ended only in 1521.

Copernicus never married and is not known to have had children, but from at least 1531 until 1539 he co-habited with Anna Schilling, a live-in housekeeper. This was seen as scandalous by two bishops of Ermland who urged him over the years to break off relations with his 'mistress'.

Towards the end of 1542 Copernicus had a stroke which paralysed him. He died on 24 May 1543 just after the publication of *De Revolutionibus*. According to legend he was shown a copy of the book on his deathbed.

3.3 SCIENCE

In Copernicus' day, the received wisdom for astronomy came from Ptolemy's *Almagest* which was written in the first century AD. The Almagest is the most thorough treatment of astronomy that appeared in the ancient world, and it contained a lot of valuable information. Unfortunately it also stated that the earth is stationary with the heavenly bodies moving around it. It's a pity that Ptolemy's statement came to be accepted because the heliocentric, theory had been proposed by Aristarchus of Samos three centuries before the time of Ptolemy. That was mentioned by Archimedes in *The Sand Reckoner* where he attributes the following words to Aristarchus

> You are now aware ['you' being King Gelon] that the 'Universe' is the name given by most astronomers to the sphere the centre of which is the centre of the earth, while its radius is equal to the straight line between the centre of the sun and the centre of the earth. This is the common account as you have heard from astronomers. But Aristarchus has brought out a book consisting of certain hypotheses wherein it appears, as a consequence of the assumptions made, that the universe is many times greater than the 'Universe' just mentioned. His hypotheses are that the fixed stars and the sun remain unmoved, that the earth revolves about the sun on the circumference of a circle, the sun lying in the middle of the orbit, and that the sphere of the fixed stars, situated

> about the same centre as the sun, is so great that the circle in which he supposes the earth to revolve bears such a proportion to the distance of the fixed stars as the centre of the sphere bears to its surface.

Copernicus seems to have proposed the heliocentric model without realising that Aristarchus got there first.

Copernicus wrote his first account of the heliocentric theory at some time before 1514. In later transcripts it's called *Nicolai Copernici de hypothesibus motuum coelestium a se constitutis commentariolus* and I'll call it *Commentariolus*. It was a brief account without mathematical apparatus, and he never intended to have it printed, instead just giving copies of it to a few people. *Commentariolus* would appear complete in print for the first time only in 1878.

Commentariolus was favourably received by the Catholic Church. In 1533, Johann Albrecht Widmannstetter delivered a series of lectures in Rome outlining Copernicus' theory. Pope Clement VII and several Catholic cardinals heard the lectures and were interested in the theory. On 1 November 1536, Nikolaus von Schönberg, Archbishop of Capua and since the preceding year a cardinal, wrote to Copernicus from Rome and asked him for a copy of his writings 'at the earliest possible moment'.

A German Protestant called George Joachim now enters the story. He was Professor of Mathematics at the recently founded University of Wittenberg. He was 25 years old and had adopted the name Rheticus because he came from the Austrian Tyrol which the Romans had called Rhaetia. Having learned of the Copernican theory he made the long, and in those times dangerous, journey to Frombork. He arrived in 1539 and stayed for two years, despite the fact that Protestants were forbidden to live in Copernicus' Catholic diocese, and despite the fact that they were being actively persecuted. Rheticus gave Copernicus several scientific books, including an edition of Euclid's geometry translated into Latin directly from the Greek. (Earlier Latin versions had been translated from Arabic.) At the same time, Rheticus had sight of Copernicus' draft of *De Revoluionibus Orbium Celestium*. With Copernicus' permission he wrote a partial summary of Copernicus' draft, called *Narratio Prima* (First Account). It was so-called because he planned to write more, but he never did.

Perhaps inspired by Rheticus' visit, Copernicus now decided to publish his great work. Copernicus had completed the manuscript by this time, and had given it for safe keeping to Bishop Tiedemann Giese. Giese in turn gave it to Rheticus. Rheticus, though, was too busy to deal with it, and he passed it on to a local Lutherian Clergyman called Andrew Osiander. That was unfortunate because Osiander wrote an unsigned preface to the publication, describing the heliocentric theory as a mathematical device to make calculations easy, which was not supposed to represent physical reality. The reader would of course assume that the preface was written by Copernicus. Anyway the book was at least published, coming out in the spring of 1543 just before the death of Copernicus. As submitted it had no title but the publishers decided to call it *De Revolutionibus Orbium Celestium Libri VI* (*Six Books on the Revolution of the Heavenly Spheres*), which I'll shorten to *De Revolutionibus.*

De Revolutionibus is regarded as one of the greatest scientific works ever written. It was written in Latin, as were nearly all European academic works before the nineteenth century. That meant that they could be read by all educated Europeans. *De Revolutionibus* stated the heliocentric theory *and* it showed in detail that the theory agrees with observation. The agreement was demonstrated for each of the planets, one by one. To account for the fact that the orbits are not exactly spherical, Copernicus started for each orbit with a point that *was* moving around the sun in a circle. Then, around that point, he drew a smaller circle of suitable size, and supposed that the planet was going around that smaller circle, called an epicycle. Epicycles had been used in astronomy since the ancient Greeks, but not in the way that Copernicus used them.

The heliocentric theory was taught in some universities in the 1500s but had not permeated the academic world until approximately 1600. It attracted a lot of criticism when it first appeared. In 1539, Martin Luther wrote this

> People gave ear to an upstart astrologer who strove to show that the earth revolves, not the heavens or the firmament, the sun and the moon ... This fool wishes to reverse the entire science of astronomy; but sacred Scripture tells us [Joshua 10:13] that Joshua commanded the sun to stand still, and not the earth.

Other people, among whom John Donne and William Shakespeare were the most influential, also feared the heliocentric theory, feeling that it destroyed hierarchal natural order which would in turn destroy social order and bring about chaos. In 1616, the Catholic Church placed *De*

Revolutionibus on the *Index of Prohibited Books*, where it remained for about 200 years.

In addition to being the founder of the Copernican Revolution, Copernicus was one of the first economists. In 1517 he stated what is now called the quantity theory of money. In 1526 he wrote *Monetae cudendae ratio* in which he stated that debased coinage (coins which don't contain the amount of silver or gold which they should) drives undebased coinage out of circulation. That's now called Gresham's law, after the English financier Thomas Gresham who made the statement several decades later.

For accounts of Copernicus and his work, see [17]–[20]. For the Protestant areas of Europe see [21]. For *De Revolutionibus* see [22]–[24]. For *Ptolemy's Almagest*, see [25].

CHAPTER 4

Kepler

Without proper experiments I conclude nothing. (Johannes Kepler in Astronomi Opera Omnia, published in 1630.*)*

4.1 BIOGRAPHY

Kepler was born on 27 December 1571 in the Free Imperial City of Weil der Stadt in the Duchy of Württemberg. He died at the age of 58 on 15 November 1630 in the Free Imperial City of Regensburg. Both places were in the *Heiliges Römisches Reich* (Holy Roman Empire of the German People) which is usually called simply the Holy Roman Empire. The Holy Roman Empire was a large territory centred on what what is now Germany. Those particular places were in what is now southern Germany.

Johannes Kepler

Kepler's paternal grandfather, Sebald Kepler, had been Lord Mayor of the city but the family fortunes had plummeted by the time that Johannes was born. His father Heinrich Kepler earned a precarious living as a mercenary, and he left the family when Johannes was five years old. He was believed to have died fighting in the Netherlands. His mother Katharina Guldenmann, an innkeeper's daughter, was a healer and herbalist who was later to be accused of witchcraft. Johannes was born two months prematurely, and he always said that he had been weak and sickly as a child. Be that as it may, he certainly astonished travelers at his maternal grandfather's inn with his phenomenal mathematical ability.

He was introduced to astronomy at an early age and developed a love for it that would span his entire life. When he was six his mother took him up a hill to see the Great Comet of 1577. When he was nine he saw an eclipse of the moon. Unfortunately, he contracted smallpox as a child which left him with weak vision and crippled hands, so that it was difficult for him to make astronomical observations as an adult.

When he was seven, Kepler went to the Latin school at Leonberg. He didn't stay at Leonberg though, because the family kept moving about. As a result his time at school, which should have been three years, stretched to five.

Kepler's parents were Lutherans. As Kepler was religious he embarked on a course of study that would lead to him becoming a minister. At the age of twelve he gained entrance to the convent grammar school of Adelberg, and two years later he entered the seminary at Maulbronn.

In 1589, when he was eighteen, Kepler entered the University of Tübingen which was a famous centre for Protestant theological studies. There he studied for five years, and there he encountered the heliocentric theory.

All of the places that Kepler had lived in so far, were in present-day Germany. In 1594 the University authorities were asked to recommend a teacher of mathematics in the Protestant school at Graz. Graz was in Austria, which at the time was called the Archduchy of Austria. Like Kepler's previous places of residence, it was part of the Holy Roman Empire. The Protestant school had been set up by nobles for their sons, but was also attended by the sons of wealthy people generally. The authorities nominated Kepler, who agreed to go even though it meant abandoning his plan to become a minister, and even though his salary would only be three-quarters of his predecessor's salary. Upon taking up his post, he turned out to be an indifferent teacher, with an erratic delivery.

In December 1595, Kepler was introduced to Barbara Müller, a 23-year-old widow (twice over) with a young daughter, and he began courting her. Müller was heiress to the estates of her late husbands and the daughter of a successful mill owner. Her father Jobst initially opposed a marriage because of Kepler's poverty. Jobst relented but the engagement nearly fell apart while Kepler was away. In the end Protestant officials who had helped set up the match pressured the Müllers to honour their agreement.

Barbara and Johannes were married on 27 April 1597. In the first years of their marriage the Keplers had two children who died in infancy. Then, in 1602, they had a daughter (Susanna); in 1604 a son (Friedrich); and in 1607 another son (Ludwig), all of whom survived infancy.

We're getting ahead though, in this short biography of Kepler. In 1596, on a two-month leave of absence from Graz, he visited Württemberg. He saw both his grandfathers. He paid his respects to the Duke of Stuttgart, who had provided a scholarship when he left home. Most importantly though, he consulted his old teacher, Michael Mästiln, about the layout of a book that he was preparing. The book was to be called *Mysterium Cosmographia* (*Cosmographic Mystery*), and it was to be the first published defence of the heliocentric theory.

Mästiln edited the text to make it easier to understand. Then, with the support of Mästiln, Kepler received permission from the Tübingen University Senate for the book to be published in Tübingen, subject to some modifications. It was published in 1596. Kepler received his copies, and he began sending them to prominent astronomers and patrons early in 1597. It wasn't widely read, but it did establish Kepler's reputation as an astronomer. The effusive dedications to powerful people, also did Kepler no harm.

One of those to whom Kepler sent a copy was Galileo. Kepler was hoping for a future collaboration, but their personalities were too different for that to happen. He also sent the book to Tycho Brahe, a wealthy nobleman and Denmark's leading astronomer. As we will see, that did lead to a fruitful collaboration.

In 1598 things seemed to be looking bad for Kepler. Although most people in Graz were Protestants, the Archduke of Württemberg and some other important people were Catholics. On 23 September 1598 the archduke ordered the expulsion of all Protestant preachers and rectors, as well as of 'school employees' which included Kepler. He departed, but he left behind his family because he hoped to be allowed to return. Fortunately, that indeed came to pass and he returned at the end of October.

In December 1599 Tycho invited Kepler to visit him in Prague. Tycho had moved there because he had fallen out of favour with the Danish Royal Court, and he was now the Imperial Mathematician of the Holy Roman Empire. Kepler had in fact already set off before receiving the invitation. He was hoping that Tycho's patronage could solve at once his scientific, social and financial problems.

He had set off in January 1600 with the retinue of an imperial emissary. On 4 February he met Tycho Brahe and two of his assistants.

They met at Benatky nad Jizerou, 35 kilometres from Prague and the place where Tycho's new observatory was being constructed. He stayed as a guest for two months, and then returned to Graz. On 10 July, he saw in the marketplace a total eclipse of the Sun.

On 27 July there was another crackdown on Protestants, and this time Kepler didn't escape it. On that date, everyone except some of the old gentry and nobility was ordered to go at a specified date to the church. There they were ordered, one by one, to become Catholics if they were not Catholics already. Kepler refused and he was ordered to leave with six months' pay. He loaded his household goods into two wagons and departed. He arrived in Prague with his family on 19 October, in poor health and with no employment.

When Kepler got to Prague, Tycho's instruments were arriving from Denmark to be set up again in a palace outside the city. Through most of 1601 Kepler was supported by Tycho. Tycho asked him to analyze planetary observations, and to write a tract against Tycho's (by then deceased) rival, Ursus.

Tycho had arranged to make Kepler a paid collaborator on a new project that he'd proposed to the emperor: it was to produce accurate tables of the positions of stars and planets that would replace the existing and quite inaccurate tables. The tables were eventually produced and were called the Rudolphine Tables in honour of the Holy Roman Emperor Rudolph II.

Kepler's interaction with Tycho wasn't without some problems. In April, when he tried to arrange a formal employment contract with Tycho, negotiations broke down in anger, but they were soon reconciled and reached an agreement on salary and living arrangements. In June, Kepler returned home to Graz to collect his family. Kepler's employment was to begin in September.

Political and religious difficulties in Graz prevented Kepler from returning immediately to Tycho. He therefore sought an appointment as a mathematician to Archduke Ferdinand, composing an essay dedicated to Ferdinand which was called *In Terra inest virtus, quae Lunam ciet* (*There is a force in the earth which causes the moon to move*). The essay didn't get the desired result.

Kepler eventually returned to Prague in September 1601. A couple of months later, Tycho was taken ill while dining at a nobleman's table and he died soon afterwards. On his deathbed he bequeathed to Kepler his instruments and his catalogue of observations. Kepler didn't immediately get them, because Tycho's heirs didn't give them to him, but he got them eventually.

Kepler was now appointed as Tycho's successor, to be the Imperial Mathematician with the responsibility to complete Tycho's unfinished work. Kepler's position in the imperial court allowed him to practice his Lutheran faith unhindered, even though that was officially not allowed. There was a financial problem though; the payment of his salary was quite irregular because of the emperor's mismanagement and because of the cost of fighting the Turks. Home life was difficult too, marred by bickering with Barbara and bouts of sickness. Court life was better because it brought Kepler into contact with other prominent scholars.

Kepler's primary obligation as Imperial Mathematician was to provide astrological advice to the emperor. Astrology is the analysis of the effect that the positions in the sky of the heavenly bodies are supposed to have on people's lives. The first step in finding the effect, is to cast a horoscope, which is a determination of the position of the heavenly bodies in the sky at a person's birth or at some other point in their life. The horoscope is then interpreted using some loosely defined rules. In Kepler's time, and for a long time after, astrology was accepted by most people and the casting of a horoscope was an important source of income for astronomers like Kepler. He had been casting well-received horoscopes for friends, family, and patrons since his time as a student in Tübingen.

In addition to horoscopes for allies and foreign leaders, the emperor sought Kepler's advice in times of political trouble. Rudolph was actively interested in the work of many of his court scholars (including numerous alchemists) and kept up with Kepler's work in physical astronomy as well.

The year 1611 was hard for Kepler's family. First his wife contracted Hungarian spotted fever. She recovered, but as she was doing so three children all fell sick with smallpox and Friedrich, age six, died.

Following his son's death, Kepler sent letters to possible patrons in Württemberg and Padua, because he wasn't happy with his situation in Prague. Concerns over Kepler's supposed Calvinistic heresies prevented his return to the University of Tübingen in Württemberg. The University of Padua, on the recommendation of the departing Galileo, offered Kepler the mathematics professorship. Kepler didn't accept because he wanted to keep his family in German territory. Instead of accepting the offer, Kepler travelled to Linz on the river Danube. Linz, like Graz, was in Austria. In Linz he obtained a position as teacher and district mathematician. The post was similar to the one that he had had in Graz. Linz, though, was something of a backwater compared with Graz, which meant that Kepler's life was less rewarding.

Kepler's wife Barbara died soon after he returned to Prague. He postponed the move to Linz, and remained in Prague until Rudolph's death in early 1612. He did no research, which isn't surprising in view of his recent tragedy and the fact that there was a dispute over his wife's estate. When Matthias became the next Holy Roman Emperor he re-affirmed Kepler's position as the Imperial Mathematician, while still letting him move to Linz.

Kepler took up his post in Linz in 1612, and he was to stay in Linz for fourteen years, the longest stay of his life in any place. His responsibilities were teaching at the district school and providing astrological and astronomical services. In Linz, he enjoyed more financial security and religious freedom than in Prague. On the downside though, his Lutheran church wouldn't allow him to take communion, because of some theological scruples that he expressed.

On 30 October 1613, Kepler married the 24-year-old Susanna Reuttinger who was an orphan of humble birth. Following the death of his first wife Barbara, Kepler had considered eleven different matches over two years! He wrote that Reuttinger (the fifth match) 'won me over with love, humble loyalty, economy of household, diligence, and the love she gave the stepchildren'. The first three children of this marriage died in childhood, but three more survived into adulthood. From what we know, it seems that this was a much happier marriage than his first. Kepler was a devoted father, and he prepared German versions of Latin texts for the use of his son Louis.

In 1615, a train of events began which could have resulted in execution of Kepler's mother Katharina. Ursula Reingold, a woman in a financial dispute with Kepler's brother Christoph, vindictively claimed that Katharina had made her fall sick by witchcraft. Katharina was arrested, but Kepler successfully defended her, and took her away to Linz in December 1616. She returned though in the summer of 1620, and in August 1620 she was imprisoned for fourteen months. She was subjected to territio verbalis, a graphic description of the torture awaiting her, in a final attempt to make her confess. She didn't confess and she was released in October 1621, thanks again (at least in part) to Kepler's defence.

In 1625 Kepler's own household ran into problems. It was decreed that all Protestants were to be expelled from Linz. An exception was made for Kepler, and for the printer who was producing his Rudolphine Tables, but Kepler's library was impounded over a dispute about the education of his children until a Jesuit friend and mathematician came to its rescue. Then the city was besieged by peasants, who were in revolt for a variety of reasons. As a result the press and the printed sheets were destroyed in a fire though the manuscript survived. When the city was relieved by Imperial forces, Kepler transferred the work to Ulm, which was farther up the Danube and in present-day Germany. He went there with his family late in 1626.

That journey was not without incident. Kepler and his family had planned to sail to Ulm but when he reached Regensburg (now Ratisbon) the river froze over. Leaving his family in that city he proceeded by wagon. Arriving in Ulm, he quarreled with the printer and decided to transfer the printing to Tübingen. He set out to walk there in the depth of winter but was forced to turn back. The book was finally printed in 1627 after Kepler overcame some obstacles put up by Tycho's heirs.

In 1628 Kepler entered into a new employment; he became an advisor to General Wallenstein, who had been a commander under Emperor Ferdinand. Kepler provided astronomical calculations for Wallenstein's astrologers and occasionally wrote horoscopes himself. During the last couple of years of his life Kepler did a lot of travelling. He went from the imperial court in Prague to Linz and Ulm, then to Sagan for a while, and finally to Regensburg.

Soon after arriving in Regensburg, Kepler fell ill, and he died on 15 November 1630. Kepler wrote his own epitaph in these words

> Mensus eram coelos, nunc terrae metior umbras
> Mens coelestis erat, corporis umbra iacet.
> (I measured the skies, now the shadows I measure
> Skybound was the mind, earthbound the body rests.)

In 1901, by the way, Kepler became a murder suspect! That was when some scientists exhumed Tycho's body and found mercury in it. They suggested that Kepler might have administered it to get his hands on the full complement of Tycho's data. He was cleared of that charge in 1910 when the body was exhumed again, and a more extensive investigation showed that Tycho hadn't ingested excessive amounts of mercury in the last five to ten years of his life.

4.2 SCIENCE: VARIOUS WORKS

We remember Kepler for his three laws, that describe the motion of planets around the Sun. Before getting on to those, I'll mention some other things.

Kepler's *Mysterium*, which appeared in 1596, was the first published defence of the Copernican system. Unfortunately, it also attempted a fruitless task, which was to explain the relative sizes of the orbits of the planets. Kepler was trying to show that sizes could be obtained by fitting together suitably shaped solids. We now know that the planets represented condensations from a hot gas that circulated the sun, and that the sizes of their orbits have no particular significance. Kepler never gave up on the search for an explanation of the sizes of the orbits. In 1621, he was to publish a second edition of *Mysterium*, half as long again as the first, setting forth things that he had worked out in the twenty-five years since its first publication. How sad that he wasted so much of his time!

In 1600, while with Tycho in Prague, Kepler analyzed some of Tycho's observations of Mars. Tycho didn't give him everything, which might have been because he didn't want to see his own earth-centred model displaced by the heliocentric model.

In 1603, Kepler worked on optical theory; the resulting manuscript, presented to the emperor on 1 January 1604, was published as *Astronomiae Pars Optica (The Optical Part of Astronomy)*. It laid the foundation for modern optics. In it Kepler states that the intensity of light falls as the inverse-square of the distance from the source. He describes the reflection of light by flat and curved mirrors, and the principles of pinhole cameras, as well as some astronomical implications of optics. He also extended his study of optics to the human eye. In 1604 he observed what we would now call a supernova, and wrote *De Stella Nova*.

Enter now Galileo. As we shall see Galileo used his new telescope in 1610 to discover four of Jupiter's satellites. Upon publishing his account as *Sidereus Nuncius (Starry Messenger)* Galileo sought the opinion of Kepler, in part to bolster the credibility of his observations. Kepler responded enthusiastically with a short published reply *Dissertatio cum Nuncio Sidereo (Conversation with the Starry Messenger)*. He endorsed Galileo's observations and offered a range of speculations about the meaning and implications of Galileo's discoveries and telescopic methods, for astronomy and optics as well as for cosmology and astrology.

Kepler also investigated the optics of telescopes himself, using a telescope borrowed from Duke Ernest of Cologne. The resulting manuscript was published as *Dioptrice* in 1611. He also observed the moons of Jupiter, producing *Narratio de Jovis Satellitibus*.

Taking time out from this serious stuff, Kepler wrote some science fiction. In 1611 he circulated a manuscript of what would eventually be published posthumously as *Somnium (The Dream)*. Part of the purpose of Somnium was to describe what practicing astronomy would be like from the perspective of another planet, to show the feasibility of the heliocentric theory. Included in *Somnium*, is a story describing a trip to the moon, and a distorted version of that story may have been in the mind of the prosecutors during the witchcraft trial of his mother. That's because the mother of the narrator consults a demon to learn the means of space travel. After his mother's acquittal, Kepler composed 223 footnotes to the story. The footnotes were several times longer than the actual text, and as well as fiction they dealt with such things as the geography of the moon.

In 1611 Kepler also composed, for his friend and some-time patron Baron Wackher von Wackhenfels, a short pamphlet entitled *Strena Seu de Nive Sexangula (A New Year's Gift of Hexagonal Snow)*. In it was the first description of the hexagonal symmetry of snowflakes. Supposing correctly that the snowflakes are made out of tiny objects (which we now call molecules) Kepler made what is called the Kepler conjecture. This is a statement about the most efficient arrangement for packing spheres. A proof of the conjecture was given in 2017 so I suppose we'll start to call it a theorem at some point.

In 1613, Kepler wrote an influential mathematical treatise, *Nova stereometria doliorum vinariorum*, on measuring the volume of containers such as wine barrels. The idea was to divide the container into horizontal circular slices. To a good approximation, the volume of each slice would be the area of the circle times the thickness of the slice.

In 1619, Kepler published *Harmony of the Universe*. It addresses the physical theory of music and contains some new findings in geometry. It also tries to relate the orbits of the planets to the theory of music.

In 1623 Kepler finally completed the Rudolphine Tables, but he encountered problems when he tried to get them published. Tycho's heirs tried to curtail his freedom to publish them as he thought best, and the Emperor had to appoint referees to adjudicate and, if necessary, to impose a settlement. There were also difficulties in the way of securing a grant to pay for the publication. Kepler went to Vienna, hoping to recover an unpaid portion of the salary that he had had when employed as the Imperial Mathematician. It's not clear whether or not he succeeded. Then there was difficulty in finding a printer. The Emperor insisted that the book be published in Linz (or anyway, in Austria) which meant that Kepler had to try to build up the local, ill-equipped press, with types and workmen brought in from outside.

The Rudolphine Tables were Kepler's best-known work during his lifetime. In addition to the actual tables there was some instruction in relevant mathematics, a chronology of world history, and a folding map of the world with a list of cities and their longitudes (with zero taken to be the longitude of Tycho's observatory). Kepler had done the vast amount of arithmetic involved without the aid of logarithms, which came too late to be of use.

As well as the *Rudolphine Tables*, Kepler produced *Ephemerides* which gave the positions in the sky of the Sun, Moon and planets for the period 1617–1636. Kepler also published astrological calendars, which were very popular and helped offset the costs of producing his other work. The offset became important when support from the Imperial treasury was withheld. In his six calendars, published between 1617 and 1624, Kepler forecasts planetary positions and weather as well as political events. The latter forecasts were pretty accurate because Kepler was well informed. They got Kepler into political trouble though, and his final calendar was publicly burned in Graz. Neither the first book burning, nor the last!

4.3 SCIENCE: KEPLER'S LAWS

What we call Kepler's laws came out bit by bit, and weren't at the time numbered or called laws. Here they are

1. Each planet moves in an ellipse with the sun at one of its two focuses.

2. A straight line from the sun to a planet sweeps out equal areas in equal times.

3. The square of the time taken to complete an orbit is proportional to the cube of the planet's maximum distance from the sun.

The second and third laws are simple enough. To understand the first law, you have to know what an ellipse is, and a focus of an ellipse. To understand that, you can do the following either in reality or in imagination. On a sheet of white paper put two pegs in the middle a short distance apart. Put a loop of string over the pegs, big enough to allow some slack. Put a pencil inside the loop and stretch it out. Finally, move the pencil all the way around the pegs keeping it always tight. The line that the pencil traces is an ellipse and each of the pegs is a focus of the ellipse. I hope that's useful.

How did Kepler arrive at his laws and when did he publish them? It wasn't easy. He began by analysing Tycho's observations of Mars and it took him about 40 failed attempts before he found that his first law applied to it. I think we're dealing here with a man that would never give up!

Finding that the law applied to the orbit of Mars he decided that it must apply to every planet, but he didn't have time to verify that, nor was he employing assistants who could have done the work for him. Further analysis of Tycho's observations led him to the second law. The content of the first two laws appeared in 1609, with the (here shortened) title *Astronomia Nova* (*New Astronomy*).

Finding the third law was s a tougher proposition than finding the first two, because the third law compares the orbits of different planets, and because it makes a quite complicated statement about the orbits. Kepler was able to find it, only after Napier discovered logarithms in 1614. Expressed in terms of logarithms, the third law says that twice the logarithm of the time taken for a planet to complete an orbit, minus three times the logarithm of of a planet's maximum distance from the sun, is the same for every planet. In this form, the third law is easily verified; all one has to do, is plot the times against the distances on logarithmic scales. Then, according to the third law, they will lie on a straight line with slope 2/3. It was the logarithmic statement which Kepler enunciated, rather than the statement that we now call Kepler's third law. Kepler announced his third law in *Harmonices Mundi* published in 1619, and he also gave it in *Epheremides* published in 1620. The latter is actually dedicated to Napier, who had died in 1617.

Napier, by the way, might have arrived at logarithms in the following roundabout way. In 1590 James VI of Scotland (later James I of England and Ireland) went to Norway to meet his prospective bride Annie. Annie was from Denmark and had travelled to meet James VI in a ship that was supposed to go to Scotland, but a storm had diverted it to Norway. (Several supposed Scottish witches were held to be responsible for the storm and burned.) On the way back the royal party visited Tycho Brahe's observatory, and James VI's physician John Craig there saw how Tycho and his assistants were using a mathematical technique called prosthaphaeresis, which in modern language corresponds to using i times the logarithm instead of the logarithm itself. When Craig arrived in Scotland, he described this to his friend Napier and it was probably this that led Napier to logarithms.

When they first came out, Kepler's laws were ignored by some important people, such as Galileo and René Descartes. Some astronomers thought that Kepler shouldn't introduce physics into his astronomy. After Kepler's death though, predictions of his theory were tested against astronomical observations. In 1631 Pierre Gassendi observed a transit Mercury across the sun on precisely the date that Kepler predicted. This was the first observation of a transit of Mercury. In 1639 Jeremiah Horrocks observed a transit of Venus.

Kepler gave a unified account of his three laws in a book called *Epitome astronomiae Copernicanae* (*Epitome of Copernican Astronomy*). The first volume was printed in 1617, the second in 1620, and the third in 1621. This book set down all three of Kepler's laws, and it became his most influential work. It applied the three laws, not just to the motion of the planets around the sun, but to the motion of the moon around the earth, and to the movement of Jupiter's then-known satellites around Jupiter. After Kepler's death, *Epitome* became very important for the dissemination of his ideas. In the period 1630–1650, it was the most widely used astronomy textbook. People like Giovanni Alfonso Borelli and Robert Hooke began to think about Kepler's laws, in terms of an attractive force exerted by the sun on the planets. That line of thinking was drawn to completion by Isaac Newton.

For Kepler's works, see the following: *Mysterium cosmographicum* [27]; Kepler's optics [28]; Kepler on Galileo's *Sidereal Messenger* [29]; Kepler's defence of Tycho Brahe [30]; *New Astronomy* [31], *The Harmony of the World* [32]; *Epitome of Copernican Astronomy* [33]; *The six cornered snowflake* [34]; Kepler's letters [35]; *The Dream* [36]. For biographies of Kepler, see [37] and [38].

CHAPTER 5

Galileo

In questions of science, the authority of a thousand is not worth the humble reasoning of a single individual. (Galileo's third letter on sunspots, written in 1612)

5.1 BIOGRAPHY

'Eppur si muove' (still it moves). So muttered Galileo after he had been forced by the Inquisition to deny that the earth is going around the sun. By the end of his life he was as famous as any person in Europe despite the best efforts of the Church.

In Galileo's time there were dramatic advances in both mathematics and science. They didn't yet come from universities but instead from the courts of kings and princes. In mathematics, decimal fractions came into use so that one could write 1.375 instead of $1\frac{3}{8}$. Regarding advances in science, I'll be focussing on theoretical physics and astronomy, where the advances largely came from Galileo himself. Comparable advances were also made in human anatomy, in biology more generally and in chemistry.

Galileo (in full Galileo Galilei) was born in 1564 near Pisa and he died in Arcetri on the outskirts of Florence in 1642. He was the first of six children of Vincent Galilei and Julia Ammoniate. Vincent was a famous lutenist, composer and music theorist, and Galileo himself became an accomplished lutenist. His mother Julia Ammoniate would later report him, more than once, to the Roman Inquisition for failing to perform his religious duties. They don't seem to have done anything about it though.

Three of Galileo's five siblings survived infancy. The youngest, Michelangelo, became a noted lutenist and composer like his father. After their father died in 1591 Galileo was entrusted with the care of Michelangelo, which was a bit of a problem. Michelangelo wasn't able to contribute his fair share of his sister's dowries, for which he was taken to court, and he sometimes borrowed money from Galileo to support his musical activities.

Galileo Galilei

Galileo seems to have been a friendly, even gregarious, person. He mixed with a wide variety of people from churchmen and courtiers to manual workers and men of the world. That meant that he had an exceptionally wide range of knowledge. It also meant that he knew how to explain things to people in familiar language. Even in his scientific writing he preferred Italian to Latin. He wrote in a witty and often sarcastic style and he had a voluminous correspondence.

After these preliminaries I'll summarise Galileo's life from childhood onwards. As already stated, Galileo was born in Pisa. When he was eight the rest of his family moved to Florence. He joined them two years later. He was at first educated by his father who was sceptical of authority. We see that from the following statement that the father made

> It seems to me that those who rely simply on the weight of authority to prove any assertion, without searching out the arguments that support it, act absurdly.

In 1775 Galileo was sent to be educated at the Camaldolese Monastery at Vallumbrosa, 35 km southeast of Florence. Its situation, 1000 metres high and surrounded by woods that had been created by the monks, could not have been more different from the city of Florence. Galileo embraced the life and joined the order as a novice but his father didn't like that development. Seeing that Galileo had an eye infection when visiting him his father used this as an excuse to take him away. Galileo resumed his studies with Valumbrosian monks in Florence but now no longer as a novice. It seems that during lessons he was constantly questioning what was being said which earned him the nickname 'The Wrangler'.

Galileo seriously considered the priesthood, but at his father's urging he instead enrolled at the University of Pisa for a medical degree. He started the degree in 1581 when he was seventeen. Fortunately for posterity, something happened in 1581 which drew Galileo away from medicine. It happened when the court of Cosimo II, the Grand Duke of Tuscany, moved to Pisa as it did each year from Christmas to Easter. After the arrival of the court, Galileo attended (incognito) some lectures that Ostinato Ricco was giving to the court pages on Euclid's geometry. This inspired Galileo to study Euclid, and he soon took some questions to Ricco who had wide interests ranging from hydraulic engineering to cosmology.

Ricco encouraged Galileo to pursue his studies, and in the summer of 1583 Galileo brought Ricco to his father's house where the two older men became friends. Ricco told Vincent that his son preferred mathematics to medicine, and asked permission to instruct him. His father agreed, but stipulated that it should appear to be against his wishes so that Galileo would not abandon his medical studies which he was currently pursuing at home.

After returning to Pisa, Galileo studied mathematics instead of focussing on the medical degree. His father Vincent found out that he was doing that, and as a result he said that he would finance Galileo for only one more year. After the year ended Galileo was forced to withdraw without completing the course. He then lived with his mother in Florence for four years. During that time he gave private lessons in mathematical subjects. Some were given in Florence and also in Sienna. He also gave lessons to the novices at Valumbrosian.

In 1588 he obtained the position of instructor in the Academia delle Parti del Diego in Florence, teaching perspective and chiaroscuro (the treatment of light and shade). While a young teacher at the Academia, he began a lifelong friendship with the Florentine painter Cipolin who later included Galileo's lunar observations in one of his paintings.

In 1589 Galileo obtained a three-year contract as an instructor of mathematics and astronomy at the University of Pisa. That got him back to science, but Pisa was not a particularly prestigious place. There was a saying.

> In Bologna there are lovers, in Padua scholars, in Pavis soldiers
> and in Pisa friars.

Pisa's university had about 600 scholars of whom two-thirds were law students. Galileo's salary was only about 1/30 of that of the highest-paid professor of medicine. His refusal to accept the received wisdom, and his insistence on the need for experiment, set him at odds with his colleagues. His contract was not renewed, perhaps because of his attitude or perhaps because he didn't ask for it to be renewed.

Fortunately, Galileo then succeeded in becoming the Professor of Mathematics at the prestigious University of Padua. The authorities there had held the post open for four years, looking for a suitable candidate. He was appointed in 1592, at three times his previous salary, and stayed for eighteen years which was more or less the rest of his working life. To make the move Galileo had to obtain the permission of the Grand Duke of Tuscany who was now Ferdinando II. Permission was granted which the Medica family later regretted.

In Padua Galileo was at first short of money because, among other things, he had to provide a dowry for his sister. He soon prospered though, moving from a small cottage into a three-storey house, with a large garden in which he entertained guests.

Padua, being situated only twenty-five miles from the busy seaport of Venice, had a lot of foreign students as well as a stream of distinguished visitors who were entering or leaving Italy. Academic standards were very high, and the environment offered Galileo ample scope for developing his ideas. He became friends with a wealthy aristocrat called Francisco Gingelli who gave Galileo access to his library of 80,000 books. Through Gingelli, Galileo met some of the most influential men in northern Italy.

Galileo was popular with his students, who built him a huge podium out of large rough-hewn planks so that they could see him better. That podium stands today in one of the halls of the University of Padua.

Galileo's lectures dealt with Euclid's geometry every other year, and on alternate years dealt with astronomy. The astronomy lectures were mainly for medical students so that they cast could horoscopes in pursuit of astrology. He also did private tutoring, offering courses called *Fortification, Use of Sector, Cosmography, Euclid, Arithmetic, Optics, Mechanics* and *Surveying*. Each course had at most a few students, sometimes only one, most of whom were foreign. He also did teaching outside the university, so that, for example, in the summer of 1605 he was in Florence as a tutor to the young prince Cosimo de' Medica.

In 1599, at the age of thirty-five, Galileo met a twenty-one year old woman called Marina Gambini while on a trip to Venice. She became pregnant, and Galileo then moved her into a house near his own. They had altogether two daughters, Virginia born in 1600 and Livid born in 1601, and there was a son Vincent born in 1606. Galileo's position in society meant that he could hardly marry Marina Gambini, and he wasn't named as the father on his children's birth certificates. For the daughters their illegitimacy was a serious issue because it would have made it difficult for them to find husbands. Galileo therefore obtained special permission to send them at a very young age to the Convent of San Mate in Arcetri which is on a hill overlooking Florence. Galileo did later legitimise his son who became his inheritor.

In 1603 Galileo became seriously ill with a bout of arthritis, which often recurred in later years. Around 1604 he began working on astronomy in order to lecture on a nova that had appeared that year. (A nova is a star that seems to appear quite suddenly. We now know that it is in fact a pre-existing star which has acquired extra material so as to become bright enough to see.)

In 1609 Galileo heard about the invention of the telescope in Holland. Without having seen an example, he constructed a superior version of the telescope and also a microscope. The telescope was more powerful than any other that existed at the time. I think that Galileo's construction of these devices, besides showing a thorough understanding of optics, tells us too what a practical person he was.

With his telescope, Galileo observed many new things, including the mountains on the moon and the satellites of Jupiter. In 1610, Galileo described his telescopic discoveries in a book called *Sidereus Nuncius* (*The Starry Messenger*). This book espoused the heliocentric view, and it was widely read so that Galileo became well known throughout Europe. The book met with approval from people with a scientific outlook such as Kepler. It also met with opposition from more traditionally-minded people. Giusto Libri, the Professor of Aristotelian Philosophy at the University of Pisa, refused to even look through Galileo's telescope. That prompted Galileo to write the following about him

> He did not choose to see my celestial trifles while he was on earth;
> perhaps he will do so now that he has gone to heaven.

The Starry Messenger came out in May 1610. In the following month Galileo resigned his Padua position, to become Chief Mathematician at the University of Pisa (without teaching duties) and, at the same time, Mathematician and Philosopher to the Grand Duke of Tuscany Cosimo II. The Grand Dukes lived in Florence, and Galileo now moved there.

Galileo's move to Florence ended his relationship with Maria Gambini. Upon moving he took his eldest two children with him. The third, who was then just four years old, joined him after a few years.

In 1611 Galileo visited Rome, in what a contemporary described as a 'tour of triumph'. It was a triumph because, at the behest of Cardinal Robert Bellarmino, Jesuit astronomers at the Collegia Roman had confirmed Galileo's findings.

I think that Bellarmino's acceptance of Galileo's findings casts light on his part in the following notorious episode. As head of the Roman Inquisition, Bellarmino had in the previous year condemned Gordon Brunt to burning at the stake. The main charge against Brunt was out-and-out heresy. Brunt rejected Christianity lock, stock and barrel, saying that he didn't believe in the divinity of Christ, nor in the existence of hell. Regarding an afterlife, he went for the reincarnation option. In addition to the charge of heresy Brunt was charged with saying that there could be planets around some of the stars on which there might be living creatures. Because of this last item he's widely regarded as a martyr to science, but in view of Bellarmino's acceptance of Galileo's astronomy I don't think he would have been convicted for that alone.

When Galileo came to Rome, the scientist and naturalist Frederica Cess hosted a banquet in honour of Galileo who was elected to Cess's *Academia dei Lincei* (*Academy of the Lynxes*). This had been founded in 1603, and its aim was the pursuit of knowledge by rational means, instead of relying on authors like Aristotle as was the practice of most university professors. In Rome Galileo also met Cardinal Maffeo Barberini, who later sided with him on a controversy over floating bodies at a court dinner in Florence.

The journey back to Florence was very tiring, and Galileo was ill for some weeks after it. Not long after he was again ill, for about three months.

Now I come to Galileo's famous conflict with the Church. It started one morning in 1613 when, at breakfast, a conversation took place. The conversation involved Cosimo de' Medici and his mother the Grand Duchess Christina. They began discussing the truth of Jupiter's satellites. Galileo's student Benedetto Castelli was present, and he later asked Galileo to comment on the central point of that conversation, namely the conflict between the Bible and the heliocentric theory. The reply was the famous *Letter to Grand Duchess Christina*, which circulated widely in manuscript form at the time. In it Galileo famously declared that the Bible teaches how to go to heaven, not how the heavens go.

Galileo's belief in the heliocentric theory alarmed Dominicans such as Tommaso Caccini and Niccolo Lorini. As a result, the Roman

Inquisition examined Galileo's letter to Christina and they didn't like it.

Apart from a general aversion to new ideas, their opposition to the heliocentric theory arose from biblical references such as Psalm 93:1, 96:10, and 1 Chronicles 16:30 which include text stating that 'the world is firmly established, it cannot be moved'. In the same manner, Psalm 104:5 says 'the Lord set the earth on its foundations; it can never be moved'. Further, Ecclesiastes 1:5 says 'And the sun rises and sets and returns to its place'.

After inspecting Galileo's letter the Inquisition declared in February 1616 that (i) the immobility of the Sun at the centre of the universe was absurd in philosophy and formally heretical and (ii) the mobility of Earth was absurd in philosophy and at least erroneous in theology. At the order of the Pope, Galileo was then summoned by Bellarmino, and was cautioned against speaking out on behalf of the heliocentric theory.

Although Galileo had only received a caution, rumours quickly began to circulate that he had been tried and condemned. Galileo therefore secured from Bellarmino a letter, stating that this was not so but that he *had* been notified of the Papal decision to censor Copernicus' *De Revolutionibus* since a heliocentric claim was contrary to the literal meaning of Scripture.

Following Bellarmino's caution Galileo kept away from writing on cosmological matters. He wrote *Discourse on the Tides* which was incorrect because it ignored the effect of the moon. In 1623, Galileo published *The Assayer* which advocated scientific enquiry as the means of gaining knowledge. In it, he wrote

> Philosophy is written in this grand book–the universe–which stands continuously open to our gaze. But the book cannot be understood unless one first learns to comprehend the language and interpret the characters in which it is written. It is written in the language of mathematics, and its characters are triangles, circles, and other geometrical figures, without which it is humanly impossible to understand a single word of it; without these one is wandering about in a dark labyrinth. (translated by Thomas Salusbury in 1661, and quoted by Machamer in *The Cambridge Companion to Galileo*, pp.64f.)

His sympathizer and patron Barberini had just been elected Pope, as Urban VIII. In 1624 Galileo had an audience with the Pope, who favourably received *The Assayer*. Galileo believed that, in the meetings he had with the Pope, he had been encouraged to discuss the heliocentric theory, so long as it was treated as an hypothesis. He therefore began to compose *Dialogue on the Two Chief World Systems*, which was published in 1632 and dedicated to the Grand Duke.

Following Galileo's agreement with the Pope, the document was put forward, not as the truth, but as a demonstration that the arguments in favour of the heliocentric theory are understood in Rome just as well as they are understood by proponents of the heliocentric theory. In other words, that the Pope's decision to ban *De Revolutionibus* was made, not because he was ill-informed, but because he held the Aristotelian view despite the arguments against it.

Dialogue on the Two Chief World Systems caused a furore because, despite Galileo's disclaimer, it was regarded as a presentation of the physical truth of the heliocentric theory. The sale of the book was suspended six months after its publication. Worse, Galileo was in September 1632 summoned to the Roman Inquisition, at which he appeared in January 1633.

First the inquisitors tried to get Galileo to say that he had earlier been officially banned from teaching Copernicus' theory as true, but Galileo produced Bellarmino's letter to contradict this. By that time, both Bellarmino and Cesi had died and Galileo, had few powerful patrons left to defend him. Galileo was then interrogated under threat of torture and made to abjure the 'vehement suspicion of heresy'. He was also forced to publicly withdraw his support for heliocentric theory and he was forbidden from publishing further work. Pleas for pardons or for medical treatment were turned down, and he was initially sentenced to life imprisonment. Eventually, after several developments, the sentence was reduced to permanent house arrest at his home in Arcetri.

Nearby, his daughter Virginia, who had taken the name Maria Celeste, was in the Convent of San Mateo. She had a close relationship with her father but she died of dysentery soon after he arrived in Arcetri. There have survived 120 letters that she sent to him. No replies have survived which probably means that they were destroyed by the Mother Superior of the convent after Virginia died.

It was around this time that Galileo uttered the famous words 'Eppur si muove' (still it moves). According to legend, he uttered them on his way out from his inquisition. There's no reason to suppose that he took such a risk, but that he did utter the words at some point is suggested by a painting dated 1643, attributed to Murillo or some painter of his school in Madrid. It shows a gaunt figure in a dungeon pointing to a wall on which the famous words are written.

At first Galileo was very upset by his sentence. It was reported that 'there were many times that he did not sleep, but went through the night crying out and rambling so crazily that it was seriously considered whether his arms should be bound'. Fortunately, he recovered his equilibrium and continued to write, producing *Discourses and Mathematical Demonstrations Relating to Two New Sciences*.

In this great work, he described all that was known about physics, which actually consisted mostly of his own findings. After the failure of attempts to publish it in France, Germany and Poland, it was finally published in Leiden in the Netherlands. The book reached Rome in 1639 and the fifty available copies quickly sold out. Galileo doesn't seem to have suffered harm from their appearance, despite the 1632 ban on his work.

By this time Galileo was going blind. He described the onset of his blindness in a letter to his friend Élie Diodati, a member of a leading Calvinist family in Geneva who had moved to Lucca in Italy

> I have been in my bed for five weeks oppressed with weakness and other infirmities from which my age, seventy four years, permits me not to hope release. Added to this (proh dolor! [O misery!]) the sight of my right eye–that eye whose labors (dare I say it) have had such glorious results–is for ever lost. That of the left which was and is imperfect is rendered null by continual weeping.

In a second letter he says that he's actually gone blind

> Alas! Your dear friend and servant Galileo has been for the last month hopelessly blind; so that this heaven, this earth, this universe, which I by my marvelous discoveries and clear demonstrations had enlarged a hundred thousand times beyond the belief of the wise men of bygone ages, henceforward for me is shrunk into such a small space as is filled by my own bodily sensations.

Galileo continued to receive visitors until, after suffering fever and heart palpitations, he died in 1642 at the age of 77. He was buried in the Basilica of Santa Croce in Florence, but not in the main body of the church where his family were buried because of his conviction for heresy. He was though reburied in the main body of the church in 1737 after a monument had been erected there in his honour.

5.2 SCIENCE

Before getting to Galileo's most important work I'll mention some other things. You can skip that if you like because it won't affect anything which follows.

From the beginning Galileo took nothing for granted. The received wisdom, for science and much else, came from the works of Aristotle who lived in the fourth century BC, but Galileo was sceptical of the Aristotelian approach. For instance, Aristotle had declared that large objects fell more rapidly than small ones, but Galileo noticed that at the start of a hailstorm, hailstones of different sizes arrived at the same time. That meant that they had all fallen at the same speed.

While living with his mother, Galileo built in 1585 an improved thermoscope. (The thermoscope was a forerunner of the thermometer, which lacked a scale.) Then, in 1586, he wrote a short treatise in Italian, *La Bilancetta* which means *The Little Balance*, that circulated in manuscript form. It opens with a short critical essay in the history of ancient science. It is this essay that contains the comment on the golden crown story, that I mentioned in the chapter on Archimedes. Then it describes La Bilancetta itself, which was an improved version of device then in use to determine the density of a liquid. It was similar to the modern Westphal balance. This treatise brought Galileo to the attention of the scholarly world.

By 1592 Galileo had completed a Latin treatise called *De motu Antiquiora* (*Older writings on motion*). It gives a description of the motion of objects, either freely falling or subject to an additional force. The description's only partially correct but it was certainly an advance on previous thinking. In 1594 he obtained a patent for a machine driven by horses, which raised water using Archimedes' screw. In 1596 he wrote a treatise on measuring heights and distance by sighting and triangulation, which survives as an appendix to a book published later.

In 1597 he produced a surveying instrument that could also be used as a calculator. About forty copies of the instrument were sold. In 1600 he wrote a treatise *De Meccaniche* (*On Mechanics*). This is an expanded version of something that he had written in 1593. It is still not entirely correct, but more correct than *De Motu.*

In 1602 he performed experiments with pendulums. He found that the time for a small swing, starting, with the pendulum at a given angle with the vertical, depends only on the length of the pendulum and not on the weight of the bob at its end. These experiments are described in a letter to Guidobalo del Monte, who was a mathematician as well as a philosopher and astronomer.

In 1604 Galileo returned to an earlier study of the motion of falling objects. To dilute the effect of gravity he studied the motion of an object along a slightly inclined plane. To be precise, he studied the motion of a bronze ball rolling down a groove. He found that the acceleration of the ball doesn't change with time and that it doesn't depend on the size of the ball.

Around the beginning of 1607 he produced an improved version of the thermoscope. He also calculated the shape of a tapered beam which, when supported at the thick end, would have equal resistance to breakage at every point. In 1609 he showed that a projectile has what is called a parabolic trajectory.

In Florence in 1611 Galileo had a discussion with his friend Salviati and two professors of the University of Pisa. It concerned the ability of objects to float in water. Galileo said, correctly, that an object less dense than water would, if immersed, float to the top, while a more dense object would sink. Crucially, he said that this didn't depend on the shape of the object. One of the professors pointed to the case of ice which floats on water. The professor said that this is because the flat shape of a sheet of ice means that it cannot sink if it is horizontal. Galileo pointed out that the sheet will rise to the surface if immersed regardless of its orientation. Then one of the professors pointed out that a sword struck flat onto water meets with much more resistance than if it is struck edge-on. Galileo said that this observation is irrelevant to the question being discussed. The sword will sink regardless of its orientation because it is more dense than water.

Shortly afterwards Galileo was told of experiments by the philosopher and poet Ludovico delle Colombe, showing that the shape of an object placed *on the surface* of water did affect its ability to float. A flat object can indeed float if placed horizontally on the surface even if it is denser than the water. This happens because of what we now call surface tension. Again, Galileo pointed out that this is irrelevant to the discussion of what happens to an object if it is immersed.

In about 1613 Galileo determined the density of air. He did this by compressing some air and measuring the downward force upon it, taking into account the upward force given by Archimedes' Principle. Also in 1613, he published *Letters on Sunspots* which describes and interprets his observation of sunspots. (These are small dark patches on the Sun. They were described in a Chinese document in 800 BC. They were also occasionally reported in medieval Europe, but were thought to be planets passing across the face of the Sun. They were studied with telescopes by two other people around the time that Galileo observed them.)

In 1616 Galileo wrote *Discourse on the Tides*. This was completely wrong, because it ignored the effect of the moon and gave only one high tide per day instead of the two that are observed. In continuing to defend his theory of the tides, Galileo's independence of thought morphed into something more like pigheadedness. He tried to argue that the two tides were local effect having something to do with the Adriatic Sea. This would mean that Lisbon, say, had only one tide per day, which Galileo thought was true until evidence was presented that it had two. Galileo never did arrive at the correct theory of the tides.

After these preliminaries, I come to the most important things that Galileo did. The first thing was to point his telescope at the sky. He did that in 1609 as soon as he had built it, and saw amazing things. He saw that the moon has mountains, and from their shadows he concluded that some of them are four miles high. He looked at the planet Jupiter, and saw for the first time that it has satellites. To be precise, he observed the four largest satellites and he determined the time that it took each one to make a complete orbit. He looked at Venus, and saw that it had phases just like our moon. (That was what made him completely sure that the planets indeed went around the sun.) He looked at the Milky Way, and saw that it consisted of individual stars. He looked at what are called nebulae, and saw that they too consisted of stars. At no time did he see a star with a discernable size, from which he concluded that stars must be far more distant than the sun. One can well imagine Galileo's excitement when making such dramatic discoveries. Within a year or so Galileo had about a hundred telescopes, of which about ten were powerful enough to observe the satellites of Jupiter.

In 1612 Galileo completed *Discourse on Bodies in or on Water*. This actually began with data on the periods of Jupiter's satellites. Then it was argued that sunspots must be on, or very near, the surface of the Sun. This is because the apparent spacing between a pair of sunspots becomes smaller as the pair moves from the centre to the edge of the Sun's disc. (In the second edition, it was noted that a sunspot, disappearing as the Sun rotated, could sometimes survive long enough to reappear when the Sun had rotated some more. It takes about a month for the Sun to do one complete rotation.) Getting on to the subject of the title, it described the dispute about floating that Galileo had had in 1611.

Another thing that Galileo did in 1612, was to write an article suggesting that an observation of Jupiter's satellites could provide an accurate determination of the time. That, combined an the observation of the sun or a star, would allow a determination of longitude at sea. An observation of Jupiter's satellites was hardly possible at sea though, and in the end the time was determined simply by using an accurate watch.

Finally, I come to Galileo's two most important works. The first of them contains the passage about observations in the cabin of a ship that I already mentioned. You might want to settle for that without bothering about the rest of Galileo's works. For those who want it though I'll summarise both works, which are presented in the form of a discussion. Throughout the discussion there is continual reference to God, particularly as the creator of everything. As the discussion is both wide-ranging and verbose, I'll just focus on what seem to me to be its most important parts.

The first work, called *Dialogue on Two Chief World Systems*, was published in 1582. It is in the form of a discussion between three men. One of them is called Salviati after a friend of Galileo. He presents Galileo's own view. The second is called Sagredo after another friend. He is initially neutral but is persuaded of the truth of Galileo's view. The third is called Simplicio. He presents the traditional Aristotelean, view. His name is fictitious being chosen presumably because Simplicio means in Italian simple-minded as well as simple.

Early in the discussion, *Dialogue* addresses the Aristotelian assertion that there are two kinds of matter; elemental of which the earth is made and celestial of which the heavenly bodies are made. The latter is supposed to be eternal and unchanging, in contrast with the former. Simplicio argues that this division is obviously well-founded because we see no change in heavenly bodies. Salviati replies that we actually do see changes in heavenly bodies, and says that if Aristotle had known about them he would have written differently. Among these changes are novas which were observed in 1572 and 1604, sunspots and comets.

Simplicio protests that if we are to abandon reliance on Aristotle, then who else should we rely on? Salviati replies that we shouldn't rely on anyone, but instead be guided by direct experience of the world.

Then there's a discussion of novas and sunspots. Simplicio feels that the novas may be just optical illusions created by the telescope. He also tries to argue that sunspots could be objects orbiting around the sun, but Salviati points that this cannot be so because all sunspots move across the sun at the same speed which means that they are on the sun's surface. Also, a sunspot appears to become smaller as it moves towards the edge of the sun, which can be explained only if it is on the sun's surface.

After some time the discussion moves on to the moon. Simplicio asks Sagredo if he believes the light parts of the moon to be water and the dark parts land. Sagredo says that this need not be so, and that he doesn't know whether it is or not. He says though, that he knows the dark parts to be very flat in contrast with the light parts. This is because the shadow of a mountain cast on a dark part has a smooth edge, whereas on a light part the edge is jagged. He says that life on the moon seems very unlikely, because the sun shines on each part of it continuously for fifteen days. Also, there can be no rain because there are no clouds. Then, the achievements of the human intellect such as geometry and the invention of writing are mentioned, and also such things as the work of Michelangelo. It is stated that marvellous as they are, they are nothing compared with the totality that is in the mind of God.

Salviati then discusses the heliocentric theory; specifically, the idea that the apparent daily rotation of the stars and planets around the sky is caused by a rotation of the earth. He points out that if instead the stars and planets were rotating most of them would have to do so with incredibly high speed, making that seem very unlikely. Salviati also clears up some misunderstanding caused by a failure to recognise that the force of gravity is everywhere directed towards the centre of the earth.

Then they consider the three main objections to the idea that the earth is rotating; that a falling object would land to the west of its original position, being left behind by the earth's motion, that a cannonball fired to the west would travel farther than one fired to the east, and that a cannonball fired vertically would land to the west. Salviati points out that these statements ignore the fact that point of release of the cannonball shares the earth's motion. He then refutes the idea that the earth cannot be moving, because we feel no motion. The crucial passage is the following, which was already mentioned in the Introduction

> Shut yourself up with some friend in the main cabin below decks
> on some large ship, and have with you there some flies, butterflies,

and other small flying animals. Have a large bowl of water with some fish in it; hang up a bottle that empties drop by drop into a wide vessel beneath it. With the ship standing still, observe carefully how the little animals fly with equal speed to all sides of the cabin. The fish swim indifferently in all directions; the drops fall into the vessel beneath; and, in throwing something to your friend, you need throw it no more strongly in one direction than another, the distances being equal; jumping with your feet together, you pass equal spaces in every direction. When you have observed all these things carefully (though doubtless when the ship is standing still everything must happen in this way), have the ship proceed with any speed you like, so long as the motion is uniform and not fluctuating this way and that. You will discover not the least change in all the effects named, nor could you tell from any of them whether the ship was moving or standing still. In jumping, you will pass on the floor the same spaces as before, nor will you make larger jumps toward the stern than toward the prow even though the ship is moving quite rapidly, despite the fact that during the time that you are in the air the floor under you will be going in a direction opposite to your jump. In throwing something to your companion, you will need no more force to get it to him whether he is in the direction of the bow or the stern, with yourself situated opposite. The droplets will fall as before into the vessel beneath without dropping toward the stern, although while the drops are in the air the ship runs many spans. The fish in their water will swim toward the front of their bowl with no more effort than toward the back, and will go with equal ease to bait placed anywhere around the edges of the bowl. Finally the butterflies and flies will continue their flights indifferently toward every side, nor will it ever happen that they are concentrated toward the stern, as if tired out from keeping up with the course of the ship, from which they will have been separated during long intervals by keeping themselves in the air. And if smoke is made by burning some incense, it will be seen going up in the form of a little cloud, remaining still and moving no more toward one side than the other. The cause of all these correspondences of effects is the fact that the ship's motion is common to all the things contained in it, and to the air also. That is why I said you should be below decks; for if this took place above in the open air, which would not follow the course of

the ship, more or less noticeable differences would be seen in some
of the effects noted. (translated by Stillman Drake, published by
University of California Press, 1953.)

Next, they discuss the Aristotelian idea that the earth is at the
centre of the universe. Salviati points out that the universe can be said
to have a centre only if it has a spherical boundary. He says that we
do not know that this is so, or even that it has any boundary at all;
it might instead be of infinite extent. Salviati then points to evidence
that the planets move around the sun, such as the observation that
Venus has phases just like the moon. He explains the motion of the
planets with the aid of a diagram.

After a while Sagredo mentions an argument against the movement
of the earth around the sun, which is that it would cause a change in
the apparent direction of the stars, which is not observed. Sagredo
says that this may be because the change is too small to be observed
with available instruments. (We now know that this was indeed the
case.) They also discuss the apparent sizes of the stars, and they say
that they do not reflect the actual sizes but rather are caused by the
behaviour of the light within the eye. In contrast the apparent sizes of
the sun and moon are their true ones, because they are much bigger
than the apparent sizes of the stars. Salviati explains how the seasons
arise from the fact that the earth's axis of rotation isn't perpendicular
to the plane of its orbit around the sun. Finally Salviati gives a long
discussion of the cause of the tides, advocating Galileo's ideas which,
as noted earlier, are not actually correct.

Now I come to Galileo's second great work. It is called (in translation) *Discourses and Mathematical Demonstrations Relating to Two
New Sciences* and was published in 1638. It is written as a discussion
involving the same three as in *The Two Chief World Systems*. Their
characters are somewhat different however. Simplicio is now represen-
tative of Galileo's earlier beliefs while Sagredo represents his middle
period and Salviati proposes his latest ideas. The two sciences referred
to are the engineering science of the strength of materials, and the
mathematical treatment of motion. The title does not, however, do
justice to the range of the book which covers Galileo's most important
scientific discoveries. As such it gives a practically complete account of
the state of physics at the time of writing. The work is divided into
four *Days*, each addressing different areas of physics.

The First Day begins with a discussion about scale. Sagredo cannot understand why with machines one cannot argue from the small to the large: 'I do not see that the properties of circles, triangles and...solid figures should change with their size'. Salviati says size obviously matters: a horse falling from a height of 3 or 4 cubits will break its bones whereas a cat falling from twice the height won't, nor will a grasshopper falling from a tower.

Then they discuss light. Since its concentrated power can melt metals Sagredo thinks that light is moving, and he describes an unsuccessful attempt to measure its speed. (As we now know the attempt was unsuccessful because light moves too quickly.) Next they discuss motion. Aristotle believed that bodies fell at a speed proportional to weight but Salviati doubts that Aristotle ever tested this. He also did not believe that motion in a vacuum was possible. Salviati asserts that motion in a vacuum is perfectly possible. He also states that in a vacuum every object would fall with exactly the same acceleration, there being no air resistance.

Measuring the speed of a fall is difficult because of the small time intervals involved. To get around this difficulty, it is proposed that one studies the motion of objects moving down a flat and almost horizontal surface. (This was the method that Galileo used.)

The Second Day begins with a discussion of balances. Then Salviati points out that animal bones have to be proportionately larger for larger animals, so that they don't break. He proves that the best place to break a stick placed upon the knee is the middle. He considers a tapering beam, which is supported at the thick end and carries a load on the other. He calculates the shape of such a beam, which will make breakage equally likely at each point. He also shows that hollow cylinders are stronger than solid ones of the same weight.

In *The Third Day*, Salviati begins with the movement of an object along a straight line, on the assumption that its speed doesn't change with time. He explains the relationship between the speed, the time since the motion began, and the distance travelled. After that he describes an experiment in which a steel ball was rolled down a groove in a piece of wooden moulding 12 cubits long (about 5.5 metres) with one of its ends raised by 1 or 2 cubits. This was repeated, measuring times by accurately weighing the amount of water that came out of a thin pipe in a jet from the bottom of a large jug of water. By this means he had been able to verify that the acceleration of the ball doesn't change with time.

In *The Fourth Day* they discuss the movement of projectiles. The motion of a projectile consists of a combination of horizontal motion with constant speed, and downward motion with constant acceleration. They show that this produces a parabolic trajectory. They show in detail how to construct the parabolas in various situations and give tables for altitude and range depending on the projected angle. Air resistance manifests itself in two ways: by affecting less dense bodies more and by offering greater resistance to faster bodies. A lead ball will fall considerably faster than an oak ball but hardly any faster than a stone ball. However the speed of a falling object stops increasing when the air resistance balances the downward force.

For further reading about Galileo see the following; Galileo's life and works [1, 2, 3] and [39]–[61], Galileo's statement about his blindness [62], *Bilancetta* [63], *De Motu* [64], *Sidereus Nuncius* [65, 66], Kepler's response to Galileo's request for his opinion of *Sidereus Nuncius* [67] Galileo's theory of the tides [68], Galileo on sunspots [69], *Dialogue on Two Chief World Systems* [70], *Discourses and Mathematical Demonstrations Relating to Two New Sciences* [71, 72]. For further reading about other things mentioned in this chapter see; thermoscope [73], measurement of longitude [74], sunspots [75], the Westphal balance [76], history of decimal fractions [77].

CHAPTER 6

Newton

Nature and nature's laws lay hid in night:/ God said, let Newton be, and all was light. (Epitaph intended for Newton, by Alexander Pope.)

6.1 BACKGROUND

6.1.1 Some history

Isaac Newton was born in 1643 and he died in 1727. During that period England, Scotland and Ireland were all ruled by the same monarch, with England and Scotland joined together as Great Britain in 1707.

The English Civil War had started the year before Newton's birth. It was caused by a disagreement as to who should control the army, which was needed to crush a rebellion in Catholic Ireland. The Royalists (Cavaliers) wanted the army to be controlled by the king, while the Parliamentarians (Roundheads) wanted it to be controlled by parliament. The war was won in 1645 by the parliamentarians led by Oliver Cromwell, and the Stuart King Charles I was executed in 1649.

The monarchy was restored in 1660 with the coronation of Charles II. He was followed in 1685 by James II who was a Catholic. His Catholic faith set him at odds with many Protestants, and 1685 he had to crush a rebellion led by his nephew the Duke of Monmouth. After that, he began to distrust Protestants and started to appoint Catholic officers to the army. He soon went further, appointing only Catholics as judges and officers of state. Also, whenever a position at Oxford or Cambridge became vacant, the king appointed a Catholic to fill it.

Knowing that James II didn't enjoy universal support, William of Orange landed in Devon in 1688 with a fleet of 500 ships containing mostly Dutch troops. He encountered no armed opposition. On 15 January 1689 the Convention Parliament met. It declared that James had abdicated and in February 1689 it offered the crown to William and his wife Mary, on the ground that Mary was the sister of James II.

William and Mary then became king and queen as equals, while James went into exile in France.

In 1690 James landed in Ireland in an attempt to regain the throne. He came with troops supplied by the Catholic King Louis XIV of France. He faced William's troops, which were Dutch, Danish, French Huguenot, German and English. James was defeated at the Battle of the Boyne and he returned to live out the rest of his life France. Legislation was now enacted which forbad the monarch to be a Catholic, or to marry a Catholic. The legislation disadvantaged Catholics, and to a lesser extent non-conformists (those who were neither members of the Church of England, nor Catholics). Newton however was a member of the Church of England.

6.1.2 Society

Before Newton's time, the only way that a scientist could tell a large number of people about a discovery had been to write a book. In Newton's time that changed, because scientific journals started to appear. Issues of the journal appeared at regular intervals, to be bought by libraries or by individuals. Then as now one submitted an article to a journal, which appeared if it was deemed acceptable. The first journals, which appeared in 1665, were the English *Philosophical Transactions of the Royal Society* (which published articles by Newton) and the French *Journal des Savants*. The former has been published continuously to the present day, and so has the latter except for an interruption during the French Revolution. Also in 1665, the first newspaper was published. It was first called the *Oxford Gazette* but soon became the *London Gazette*, which is published to this day. (Some people discount the *London Gazette* because it's an official government organ. If we discount it, the first newspapers were published in the early 18th century.)

Despite the advances in knowledge and its dissemination, Newton's era was not at all like ours. About two-thirds of men and two-thirds of women, were illiterate. Women were still being hanged for witchcraft, or burned at the stake. In 1685, following the failed Monmouth Rebellion which attempted to remove James II, 320 men were executed. Most of the executions consisted of being hung, drawn and quartered, which for many centuries had been the method of execution for a man convicted of treason. It consisted of hanging and disemboweling, and then the body being cut into quarters. (A woman convicted of treason was burned at the stake.)

I have a personal interest in treason because of something that happened in the house of one of my ancestors. Treason was, at that time, supposed to be defined as active opposition to a monarch, which it was in the case of the Monmouth Rebellion. But in practice it was interpreted far more widely. In particular, to be a Catholic priest might be regarded as treason. One priest, the eighty-two-year-old Nicholas Postgate, was arrested in the house of my great great great great great great great grandfather Matthew Lyth. Postgate had said mass in the mass house in the then-remote village of Egton Bridge on the river Esk which flows down to Whitby in Yorkshire, and when he was arrested he was baptising a baby.

He was executed in York in 1679 and was lucky in that the hanging broke his neck so that he was dead before the drawing and quartering. (That was often not the case because hanging at that time was quite inefficient.) On the other hand he was unlucky to have been arrested at all, because he'd been working as a priest in the Esk valley for forty years. The train of events leading to his arrest had started with the murder of the employer of one Simon Reeves; Catholics were blamed, and in revenge Reeves went with an accomplice to the Esk valley which was known to be a hotbed of Catholics. It was they who seized Nicholas Postgate and gave him up for trial.

6.2 BIOGRAPHY

The mathematician Lagrange wrote that Newton was the 'greatest genius who ever lived', as well as 'the most fortunate, for we cannot find more than once a system of the world to establish'. He was talking about Newton's laws of motion which describe the fall of an apple from a tree, the moon's orbit around the earth, the earth's orbit around the sun, the rotation of our galaxy and its voyage through vastness of space. Truly, Newton had a unique opportunity to understand the world and he made the most of it—when he wasn't occupied with alchemy, or studying the Bible.

For Newton was a man of his time. About ten percent of his known writing (including annotated copies of other manuscripts) is concerned with alchemy; and, using the Bible, he estimated that the world would probably end in 2060, though it might be later.

Isaac Newton

Isaac Newton was born on 4 January 1643 (the year after the death of Galileo) and he died at the age 84 on 31 March 1727. He was born in Woolsthorpe, Lincolnshire. His father had died three months earlier. The father, who had been illiterate, had owned property and land which Newton's mother would have inherited. When Newton was three years old the mother remarried. She went to live with her new husband while Isaac lived with her mother and father in Woolsthorpe. Newton doesn't seem to have had a good relationship with his mother and stepfather. When examining his sins at the age of nineteen he wrote 'threatening my father and mother Smith to burn them and the house over them'. Isaac was the only child of his mother by her first marriage but she had three children by her second marriage.

Upon the death of his stepfather in 1653 Isaac lived in an extended family consisting of his mother, his grandmother, one half-brother, and two half-sisters. From shortly after this time, at the age of twelve, Isaac began attending the King's School in Grantham. This was five miles from his home and Isaac lodged at Grantham, with a pharmacist named Clark. There he was interested in Clark's laboratory and library. He also built various devices to amuse Clark's daughter including a windmill run by a live mouse, floating lanterns and a sundial.

Newton seems always to have been adept with his hands. In later life, he was asked by John Conduitt where he obtained the tools to grind reflectors he said that he made them himself and added 'If I had staid for other people to make my tools and other things for me I had never made anything of it'. On the other hand, while at school he seems to have shown little promise in academic work. His school reports described him as 'idle' and 'inattentive'. This is perhaps surprising because we know from his notebooks that the King's School curriculum included a significant amount of mathematics, including trigonometry with the calculation of sine tables.

Isaac's mother, by now a lady of reasonable wealth and property, thought that her eldest son was the right person to manage her affairs and her estate. When he was seventeen she pulled him out of school with the intention that he should become a farmer. He plainly disliked this occupation though, and an uncle, William Ayscough, thought that he should prepare for entering university. He persuaded Isaac's mother to return him to the school. The headmaster of the school seems also to have wanted Newton to return. If so he must have been showing more promise than his school reports suggest. Another piece of evidence for Isaac's love of learning comes from his list of sins, which include '... setting my heart on money, learning, and pleasure more than Thee ...'. Upon his return to school he lived with Stokes and paid attention to his studies so that he became the top student.

During his time in Grantham he developed a close relationship with Catherine Storer, a step-daughter of Clark with whom he was lodging. By the time that he left for Cambridge, in June 1661 when he was nineteen, it was understood that they were engaged to be married. That was not feasible until Newton had an income and as the years went by the idea that they would be married faded away. Newton never did marry and is generally considered to have died a virgin. Presumably unaware of Newton's love for Catherine Storer, Voltaire wrote that 'he was never sensible to any passion, was not subject to the common frailties of mankind, nor had any commerce with women'.

At Cambridge he entered his uncle's old college, Trinity. Newton's early years at Cambridge coincided with the Restoration era. Many of his fellow students would have been from the Public Schools and would have had relatively lax morals. I don't suppose that Newton, coming from a different background, felt himself to be one of them.

Upon arriving in Cambridge he at first earned money by performing valet's duties. His mother didn't finance him even though she could easily have afforded to do so, but it's been suggested that he received help from Humphrey Babbington. Babbington was a distant relative of Newton's, and was a Fellow of Trinity. In 1664 Newton was awarded a scholarship. This would have financed him for four more years, allowing him to obtain an MA, but fate intervened.

After Newton obtained his BA in 1665 an outbreak of bubonic plague, known as the Great Plague, arrived in London and soon spread across the country. We now know that the plague was carried by the fleas that lived on rats. Unfortunately, people at the time thought that it was carried by cats and dogs, and accordingly the Lord Mayor of

London had about 40,000 dogs and 200,000 cats killed. That, of course, increased the rat population. In London the Great Plague killed about 60,000 people which was about one-fifth of the population.

To avoid the plague, Newton's college closed and he returned to his former home in Woolsthorpe. His eighteen months there were very productive. Later, in 1718, he wrote the following in the draft of a letter to Pierre Des Maizeaux, a French Huguenot living in London.

> In those days I was in the prime of my age for invention, & minded Mathematicks & Philosophy more than at any time since.

(The draft is now in the Cambridge University Library.)

While in Woolsthorpe he worked on optics, but most notably he formulated his law of gravity. It was at this time that the famous incident of the apple seems to have occurred. We have an account of it from William Stukeley, an archaeologist who was a friend of Newton and one of his first biographers. He wrote the following.

> After dinner, the weather being warm, we went into the garden and drank tea, under the shade of some apple trees...he told me, he was just in the same situation, as when formerly, the notion of gravitation came into his mind. It was occasion'd by the fall of an apple. Why should that apple always descend perpendicularly to the ground, thought he to himself. Why should it not go sideways, or upwards? But constantly to the Earth's centre? Assuredly the reason is, that the Earth draws it. There must be a drawing power in matter. And the sum of the drawing power in the matter of the Earth must be in the Earth's centre, not in any side of the Earth. ... Therefore the apple draws the Earth, as well as the Earth draws the apple.

There's no mention though of an apple falling on Newton's head, and nobody seems to know how that myth originated.

In April 1667, Newton returned to Cambridge, and was elected to a minor fellowship of Trinity College. Fellows were supposed to take a vow of celibacy, and to undertake to become ordained priests within seven years of completing the MA, but that rule was not being enforced. Newton didn't want to become a priest, and he might not have gone for the MA if the rule had been enforced. In July 1668 he was elected to a major fellowship, which allowed him to dine at the Fellow's Table.

Newton received his MA in 1668. Only one year later, at the age of twenty-seven, he became the Lucasian professor of mathematics, and he remained in that position for another twenty-seven years. Lucasian professors were so-called because the position was initially funded by a bequest from Henry Lucas, Cambridge University's first Member of Parliament. The first Lucasian Professor was Barrow, who had apparently recommended Newton for the position. His successors have included Charles Babbage, generally regarded as the father of computing, George Stokes, one of the founders of the science of fluid dynamics, Joseph Larmor whose work laid part of the foundation for Einstein's, Paul Dirac who was one of the founders of quantum field theory which describes elementary particles and Steven Hawking who was probably the world's most famous scientist of recent times.

Newton's lectures weren't popular. Humphrey Newton (no relation) recorded that 'so few went to hear Him, & fewer yet understood him. Yet oftimes he did in a manner, for want of Hearers, read to ye walls'.

In 1678 Newton had a violent exchange of letters with the English Jesuits in Liege, concerning his theory of colour. He appears to have suffered a nervous breakdown as a result. His mother died in the following year and he withdrew further into his shell, mixing as little as possible with people for a number of years.

Newton's withdrawal may have been good for the advancement of knowledge, because in 1687, at the age of forty-five, he published in Latin his great work *Philosophieae naturalis principia mathematica* (*Philosophical principals of natural philosophy*), generally known just as *Principia*. It describes Newton's laws of motion, and their consequences for the motion of the planets and the moon. Upon its publication the book was widely acclaimed as a great advance of science, and now it's generally regarded as the greatest scientific book every written. It's been translated into English and many other languages. The book made Newton famous even though few people understood it. (More than two centuries later Einstein was to be in the same position.)

The intensity of Newton's work on *Principia* is described in reminiscences from Humphrey Newton. He tells of Isaac Newton's absorption in his studies, how he sometimes forgot his food, or his sleep, or the state of his clothes, and how when he took a walk in his garden he would sometimes rush back to his room with some new thought, not even waiting to sit before beginning to write it down. Other evidence also shows Newton's absorption in *Principia*: Newton for years kept up a regular programme of chemical or alchemical experiments, and he normally kept dated notes of them, but for a period from May 1684 to April 1686, Newton's alchemy notebooks have no entries at all. So it seems that Newton dropped his other activities, working almost exclusively on *Principia* for well over a year and a half.

In 1685, when James II became king, Newton had occasion to become involved in the politics of Cambridge University. As we have seen, the policy of James II was to appoint only Catholics when a position at Oxford or Cambridge became vacant. Newton was a staunch Protestant and was strongly opposed to this policy. When the king tried to insist that a Benedictine monk be given a degree, without taking any examinations or swearing the required oaths, Newton wrote to the Vice-Chancellor 'Be courageous and steady to the Laws and you cannot fail'. Unfortunately for him, the Vice-Chancellor took Newton's advice and was dismissed from his post. Newton didn't abandon him though; he argued the case strongly, preparing documents to be used by the University in its defence.

After the publication of *Principia*, Newton became depressed and lost interest in scientific matters. In 1688 the University of Cambridge elected Newton, now famous for his strong defence of the university, as one of their two members to the Convention Parliament. Newton was at the height of his standing–seen as a leader of the university and one of the most eminent mathematicians in the world.

Newton had a close friendship with the Swiss mathematician Nicolas Fatio de Duillier, whom he met in London around 1689. Their friendship came to an abrupt and unexplained end in 1693 and at this time Newton suffered another nervous breakdown . During the breakdown he sent wild accusatory letters to his friends Samuel Pepys and John Locke. His letter to Locke said that Locke had 'endeavoured to embroil me with women'. There might be a simple explanation for Newton's erratic behaviour in later life; his hair was examined after his death, and was found to contain mercury. The mercury would have come from his work on alchemy.

In 1696, at the age of fifty-four, Newton left Cambridge to become Warden, and after three years Master, of the Royal Mint in London. With this salary added to income from his estates Newton became a very rich man. The job with the Mint was supposed to be a sinecure but Newton took it seriously. He made changes in the money system that were effective for more than a hundred years and he also took steps against the counterfeiting of coins. At that time, counterfeiting constituted the offence of high treason, punishable by hanging, drawing and quartering. In disguise Newton went to taverns to gather evidence.

He had himself made a justice of the peace and successfully prosecuted twenty-eight counterfeiters and coin clippers.

Though Newton did no research at this stage his mathematical powers remained. In 1696 the Swiss mathematician Johann Bernoulli posed a set of problems that he thought would occupy mathematicians for the next hundred years. Receiving the list one evening Newton chose one of the problems and solved it before going to bed. When Bernoulli saw the solution he easily guessed that it was by Newton, saying that 'a lion is known by its claws'.

Newton was elected president of the Royal Society in 1703, and was re-elected each year until his death. He was given a knighthood in 1705 but the award seems to have been motivated by political considerations rather than in recognition of his scientific work.

Around the turn of the century Newton became involved in a bitter controversy with Leibnitz, as to which of them had invented calculus. In his capacity as President of the Royal Society he appointed an 'impartial' committee to decide which of them was the inventor. The report appeared in 1712 and was written by Newton although it appeared anonymously. The report suggested that Leibnitz already knew about Newton's calculus before he worked on that subject himself. (It's now known that, in fact, the two of them discovered calculus independently.) Newton then wrote a review, again anonymously, making the same claim, which appeared in *Philosophical Transactions of the Royal Society*.

Mention should be made of Newton's interests other than in mathematics and physics. Newton was a committed, though rather unorthodox, Christian. His desire that *Principia* would strengthen a belief in God is expressed in the following words

> When I wrote my treatise about our Systeme I had an eye upon such Principles as might work with considering men for the beliefe of a Deity and nothing can rejoyce me more then to find it useful for that purpose.

He even went further, by giving what is sometimes called the argument from intelligent design for the existence of God. In *Observations upon the Prophecies of Daniel and the Apocalypse of St. John* he tried to show that predictions found in the Bible had come true. In *Chronology of Ancient Kingdoms*, published in full after his death, he tries to establish the dates of historical events by linking Egyptian, Greek and Hebrew histories.

As already mentioned, about ten percent of his *known* writing is on alchemy. More of his writing may have been lost in a fire in his laboratory. Alchemy was in some respects a precursor to chemistry, which in Newton's day was getting underway with the work of the Anglo-Irishman Robert Boyle. One of the goals of alchemy, in which Newton seems to have been interested, was the discovery of the Philosopher's Stone, a material which was supposed to change ordinary metals into gold. Some practices of alchemy were banned during Newton's lifetime. This was partly to protect people from swindlers and partly because it was feared that a discovery of the Philosopher's Stone would devalue the currency. Because of the ban and perhaps also because he was very sensitive to criticism, Newton didn't publish his work on alchemy and it didn't appear when he died either because his executors found it 'unfit to publish'. Some of Newton's alchemical writing was published in 1936 and most of it's now available online. ˙

Like many scientists and mathematicians, Newton seems to have been very absent-minded. One author made the following statement

> Always preoccupied with his profound researches, the great Newton showed in the ordinary affairs of life an absence of mind which has become proverbial. It is related that one day, wishing to find the number of seconds necessary for the boiling of an egg, he perceived, after waiting a minute, that he held the egg in his hand, and had placed his seconds watch (an instrument of great value on account of its mathematical precision) to boil!

In a similar vein another author wrote the following

> As to his manners, he dressed slovenly [and had] extreme absence of mind when engaged in any investigation.... On the few occasions when he sacrificed his time to entertain his friends, if he left them to get more wine of for any similar reason, he would as often as not be found after the lapse of some time be found working out a problem, oblivious alike of his expectant guests and of his errand.

When Newton was in London his half-niece Catherine Barton went to live with him. She acted as his hostess in social affairs. Catherine was beautiful and witty, attracting the admiration of such famous figures as Jonathan Swift and Voltaire. Newton was also fond of her and after she contracted smallpox he wrote 'Pray let me know by your next, how your face is and if your fevour be going. Perhaps warm milk from ye Cow may help to abate it. I am your loving Unkle, Is. Newton'. In the final years of his life, Newton lived with Catherine and her husband at Cranbury Park near Winchester.

In old age, Newton suffered from gout, inflammation of the lungs and a painful kidney stone. He died on 31 March 1727, and was buried in Westminster Abbey with great ceremony. Voltaire observed that 'he was buried like a king, who had done well by his subjects'. His pall was born by two dukes, three earls and the Lord Chancellor.

Towards the end of his life Newton wrote in a letter to Robert Hooke these words. 'If I have seen further than others, it is by standing on the shoulders of giants'. He's also supposed to have written the following

> I do not know what I may appear to the world, but to myself I seem to have been only like a boy playing on the sea-shore, and diverting myself in now and then finding a smoother pebble or a prettier shell than ordinary, whilst the great ocean of truth lay all undiscovered before me.

There was indeed an undiscovered ocean, that will be explored to the end of civilisation, but it was Newton who launched the boat and guided it through the shallows into open waters.

6.3   MATHEMATICS AND OPTICS

Newton's work was made possible by developments in mathematics during the early seventeenth century. One of the most important of these was the invention of logarithms that I described in the chapter on Kepler. Without logarithms the complex calculations of Newton and of Kepler before him could not have been done. Another important development was the discovery by the French mathematician, philosopher and scientist René Descartes, of Cartesian coordinates. These allow the position of a point in space, to be represented by three numbers. They also allow any curve to be defined. Before that only certain curves were defined, usually with reference to geometry. An ellipse, for example, was defined as the line obtained by cutting the surface of a cone with a suitably oriented flat surface.

During his first two years at Cambridge, Newton attended lectures on arithmetic, Euclid's geometry and trigonometry. He also read, or listened to, lectures on the heliocentric theory. For most of his work after the Easter term of 1663 he was under the direction of Isaac Barrow, the newly appointed Lucasian Professor. Among other things, Barrow taught him optics. Newton also read works by Descartes, John Wallis and other authors.

Wallis, living from 1616 to 1703, was an English clergyman, mathematician and theoretical physicist. He was among those, whose work laid the foundation for calculus and he seems to have had extraordinary mathematical powers. Sleeping badly one night he calculated in his head the square root of a number with fifty-three digits. In the morning, he remembered and dictated the twenty-seven digit result. After reading Wallis' account of something he did in mathematics, Newton summarised it and then wrote 'Thus Wallis doth it but it may be done thus. . . '.

Not long before leaving for Cambridge Newton had begun to make notes in Latin. Arriving in Cambridge he continued with the notes. He called them *Quaestiones quaedam philosophicae* (*Questions about philosophy*) but they were never published. They are now called the Portsmouth Collection, and are in the library of Cambridge University. In *Quaestiones* Newton describes how he taught himself mathematics. To do that he read works on algebra, geometry, trigonometry and probability. The works were by various authors, but the single author who influenced Newton the most was Descartes. *Quaestiones* also describes how Newton studied optics and the heliocentric theory. He headed *Quaestiones* with a Latin statement meaning 'Plato is my friend, Aristotle is my friend, but my best friend is truth'.

In 1701 Newton laid down his Law of Cooling, which states that the rate of cooling of an object is proportional to the difference of temperature between the object and its surrounding. The law appeared as an article in *Philosophical Transactions of the Royal Society*, written anonymously in Latin with the title *Scala graduum caloris* (*Scale of the degrees of heat*).

Next we come to Newton's study of optics. He became interested in that subject after reading a treatise called *Physico-mathesis de lumine* (*The mathematical physics of light*). This had been published in 1666 by Francesco M. Grimaldi who was professor of mathematics and physics at the Jesuit college in Bologna. In it Grimaldi explains that the shadow of a thin rod, seen on a screen, does not consist simply of a dark band the size of the rod. Instead it consisted of alternate dark and light bands. This is an example of what is called the diffraction of light. To explain diffraction, Grimaldi suggested that a light beam is a wave.

Newton described his optical investigations in his *Opticks* of 1704. Except for a portion at the end this was written in English rather than in Latin, which was unusual at the time. (It was translated into Latin in 1706.) It was published long after he had done the work because Newton was waiting for the death of Robert Hooke, with whom he had sometimes bitter correspondence concerning priority of discovery. In *Opticks*, Newton explained that white light could be decomposed into primary colours by a prism, because these colours were refracted through different angles. He also showed that the white light could be restored by passing the primary colours through a different prism. He argued that all other colours consist of a mixture of the primary colours. Newton's findings overturned existing (Aristotelian) views concerning light.

These were notable discoveries, that would have made Newton's name known if he'd done nothing else. He didn't get everything right though. In *Opticks*, he stated that a light beam consists of particles, and showed that this can explain all aspects of light, except for diffraction. To explain diffraction, Newton had to suppose that the particles of light generated waves. I don't know why Newton didn't just declare that light itself is a wave!

Once Newton had found that different colours undergo different amounts of refraction, he realised that a telescope with a lens would always produce an imperfect image. He therefore built, entirely unaided, a telescope using a mirror instead of a lens. It was the world's first reflecting telescope. He donated a copy of it to the Royal Society, and was elected a Fellow. (At present, the world's largest telescope mirror is on Gran Canaria and it's 10.4 metres wide.)

In 1672 Newton sent an account of his work on optics to *Philosophical Transactions of the Royal Society*. When it was published the paper attracted several critical responses. At first Newton answered each one patiently but when this produced more criticism he became angry and vowed never to publish again, or even work on science. He did indeed confine himself to alchemy and theology for a while but fortunately he then returned to science.

Newton made many advances in mathematics. His biggest advance was the discovery of calculus. That discovery is described in a Latin manuscript *De Analysi* which, in 1669, was in the possession of John Barrow. He sent it to John Collins, a largely self-taught mathematician with a particular interest in surveying. (Collins corresponded with several leading mathematicians at various times. These included Newton himself and Leibnitz, as well as John Wallis and the Italian Giovanni Borelli who worked also in physics and physiology.) In his covering letter to Collins, Barrow described Newton as '... very young ... but of an extraordinary genius and proficiency in these things'. Later, in 1671 Newton recorded some more of this mathematical work in *De Methodis Serierum et Fluxionum* which was not published until John Colson produced an English translation in 1736.

6.4 THE *PRINCIPIA*

Finally, we come to Newton's work on mechanics which was his crowning achievement. Already in the 1660s Newton studied collisions. He found that the movement of the centre of mass of two objects isn't affected if they bounce off each other.

In January 1684, there took place a conversation between three men. One was the astronomer, geophysicist, mathematician, meteorologist and philosopher Edmond Halley who computed the orbit of Halley's comet. The second was the anatomist, astronomer, geometer, mathematician, physicist and architect Christopher Wren who rebuilt St. Paul's Cathedral after the Great Fire of London. The third was the remarkable Robert Hooke whose fields of activity included mathematics, physics, biology and mechanical inventions such as the clock pendulum. In this conversation Hooke claimed to have stated the inverse-square law of gravity and to have derived from it the laws of planetary motion. He did not produce this derivation though, even after some time, and as a result Halley reported the conversation to Newton.

Newton replied that he had derived these results himself but couldn't find the papers. In late 1684 Newton sent Halley a nine-page Latin document titled *De motu corporum in gyrum* (*Of the motion of bodies in an orbit*). In this document Newton derived Kepler's laws of planetary motion from the inverse-square law of gravity. Halley saw how remarkable it was and visited Newton again to ask him to produce more. He indeed did so publishing in 1687 *Philosophiae Naturalis Principia Mathematica* (*Mathematical Principles of Natural Philosophy*) (called usually simply *Principia*) which incorporated the 1684 publication. Two further editions of *Principia*, with some extensions and corrections, were published in 1713 and 1726.

As *Principia* is so important I'll give some more information about its publication. *Principia* is in three volumes. Newton called them books, and I'll use that terminology. Up to some time in 1685 Newton conceived *Principia* as two books. The first book became, in extended form, Book 1 of *Principia*. The second book was going to apply the results of the first book to the earth, the moon, the tides, the solar system and the Universe. In that respect it's similar to Book 3 of *Principia* but is written in a far more readable style. Newton explained that the change of style for *Principia* was made so that Book 3 could be read only by those who had mastered the previous books.

After Newton's death in 1727 the relatively accessible character of Newton's originally planned second book led to the publication of an English translation in 1728. It appeared under the English title *A Treatise of the System of the World*. Newton's heirs shortly afterwards published the Latin version in their possession under the title *De Mundi Systemate*.

While writing *Principia*, Newton had exchanged a few letters with the astronomer John Flamsteed about observational data on the planets. He acknowledged Flamsteed's contributions in the 1687 version of *Principia*.

The text of the first of the three books was presented to the Royal Society at the close of April 1686. Hooke then made some priority claims without substantiating them. When Hooke's claim was made known to Newton, who hated disputes, he threatened to withdraw and suppress Book 3 altogether. Fortunately for posterity, Halley was able to persuade Newton to let it go forward. Samuel Pepys, as President of the Royal Society, gave his imprimatur on 30 June 1686, licensing *Principia* for publication. The Society had just spent its book budget on a History of Fishes and the cost of publication was borne by Edmund Halley. The first edition of *Principia* appeared in summer 1687. Since only between 250 and 400 copies were printed by the Royal Society the first edition is very rare. In 2016 a copy was sold for $3.7 million.

The second edition of Principia came out in 1713. Several people had given Newton corrections to the first edition. One of them was the French physicist, philosopher and theologian Firmin Abauzit. A second was the English mathematician Roger Cotes. A third was the Scottish mathematician and astronomer David Gregory, and a fourth was Flamsteed. Unfortunately Newton failed to acknowledge some of these in the second edition. The third edition was published 25 March 1726, under the stewardship of the physician and man of letters Henry Pemberton. Newton described him as 'a man of the greatest skill in these matters...' and Pemberton later said that this recognition was worth more to him than his 200 guinea fee.

I now summarise the content of *Principia*. Regarding mathematics the most important item is a geometrical form of calculus. Coming to the physics one thing that's done is a calculation of the speed of sound. It is found that the speed of sound is about 979 feet per second. That's about fifteen percent lower than the true value which had already been established by experiment. The discrepancy comes from the fact that Newton ignored the effect of the temperature fluctuation, caused by the sound wave.

The most important physics in *Principia* relates to the movement of objects. It's stated that a force on an object causes an acceleration, which is equal to the force divided by the weight of the object. Newton's law of gravity is given. This is the statement that the force between two spherical objects is equal to the product of their weights, times a number called the gravitational constant, divided by the square of the distance between their centres. It's shown that from those two things one can derive Kepler's three laws for the orbit a planet around the Sun. It's also shown that one can explain the orbit of the Moon around the Earth.

In this discussion Newton recognises that, though one usually takes the Sun to be at rest, it's actually the centre of gravity of the solar system that can be taken to be at rest. The Sun is near the centre of gravity, but it's jiggled about by the pull of the planets. Newton estimates the mass ratios Sun:Jupiter and Sun:Saturn, and concludes that the amount of jiggling 'would scarcely amount to one diameter of the Sun'.

The fact that the earth's rotation makes it bulge around the equator is discussed. The slow change (precession) in the direction of the earth's axis of rotation is accounted for. Newton also gives the theory of the motions of comets, for which much of the data came from John Flamsteed and Edmond Halley. He also gives, for the first time, a correct explanation of the tides which takes into account the effect of both the sun and the moon.

In *Principia*, Newton recognises that, as far as his laws of motion go, the centre of gravity of the solar system need not be at rest. They would apply equally well if the centre of gravity were moving along a straight line with a fixed speed. Newton rejected the second alternative though, writing that 'the centre of the system of the world is immoveable', which 'is acknowledg'd by all, while some contend that the Earth, others, that the Sun is fix'd in that centre'.

Newton's rejection of the second alternative is at odds with the Principle of Relativity, because it implies that it makes sense simply to say that something is moving, as opposed to saying that it is moving relative to some other object. On the other hand Newton's statement that his laws would apply to the solar system if its centre of gravity were moving along a straight line with a fixed speed, tells us that the laws themselves are consistent with the Principle of Relativity.

Newton's view that it makes sense to say simply that something is moving, is explained in the following words

> Only I must observe, that the vulgar conceive those quantities under no other notions but from the relation they bear to perceptible objects. And it will be convenient to distinguish them into absolute and relative, true and apparent, mathematical and common. [...] instead of absolute places and motions, we use relative ones; and that without any inconvenience in common affairs; but in philosophical discussions, we ought to step back from our senses, and consider things themselves, distinct from what are only perceptible measures of them.

Regardless of these philosophical considerations, the sheer number of phenomena that could be organised by Newton's theory was so impressive that methods and language of *Principia* came to be widely adopted. Perhaps to reduce the risk of public misunderstanding, Newton in the second and third editions, gave four 'Rules of Reasoning in Philosophy'. In their final form in the third edition they are as follows

1. We are to admit no more causes of natural things than such as are both true and sufficient to explain their appearances.

2. Therefore to the same natural effects we must, as far as possible, assign the same causes.

3. The qualities of bodies, which admit neither intensification nor remission of degrees, and which are found to belong to all bodies within the reach of our experiments, are to be esteemed the universal qualities of all bodies whatsoever.

4. In experimental philosophy we are to look upon propositions inferred by general induction from phenomena as accurately or very nearly true, not withstanding any contrary hypothesis that may

be imagined, till such time as other phenomena occur, by which they may either be made more accurate, or liable to exceptions.

Newton's statement of the four rules represented a major advance in thinking.

In the third edition of *Principia* Newton explains each rule, sometimes giving an example to back up what the rule is claiming. The first rule is explained as a principle of economy. The second rule states that if one cause is assigned to a natural effect, then the same cause so far as possible must be assigned to natural effects of the same kind. Examples are respiration in both humans and in animals, fires in both the home and in the Sun, and reflection of light whether terrestrial or from the planets. An extensive explanation is given of the somewhat unintelligible third rule. In the fourth rule, Newton discusses the generalisation of observational results, with a caution against making speculations contrary to experiments.

The idea of a gravitational force acting over vast distances seemed to some people unreasonable. Newton firmly rejected such criticisms and in later editions of *Principia* he wrote that it was enough that the force explained what was observed. There was no need to enter into philosophical discussion about it.

Newton's description of the movement of objects set the pattern for all future laws of physics, that deal with changes over time. The idea is that given the state of affairs at some particular time (an initial condition) the laws of physics decide what happens subsequently. With the advent of quantum physics there was a partial retreat from that programme. In that context the initial conditions do not determine what happens next. There can be a range of possible outcomes and the role of the laws of physics is just to assign a probability to each outcome. Still, the idea is much the same.

For further reading about Newton see the following. Biographies [1], [3], [2], [78]–[86], his religion [87], the first of the above accounts of his absentmindedness [88], the second account [89], his statement 'if I had stayed for other people to make my tools ...' [79] the Portsmouth papers [90, 91], his optics [92, 93], his *Principia* [94]–[99], his science in general [100, 101], his mathematical papers [102], his philosophical writings [103], his library [104], his exchange of letters with the Jesuits [105], all of his correspondence [106], all of his writings [107], the Royal Society report of 1712 [103], Humphrey Newton's letters, including the remark about Newton's lectures [108], story of the falling apple

[109, 110]. For more about other things mentioned in this chapter, see the following; the English civil war [111], seventeenth century science [112], the *Journal des Savants* [113, 114], *Philosophical Transactions of the Royal Society* [115], the history of British newspapers [116], literacy in Britain [117], the plague [118].

CHAPTER 7

Oersted and Faraday

I could trust a fact, and always cross-examined an assertion. (quoted in The Life and Letters of Faraday *by Bence Jones)*

7.1 INTRODUCTION

In this chapter I deal with Hans Christian Oersted from Denmark who lived from 1777 until 1851, and with Michael Faraday from England who lived from 1791 until 1867. I've put them together because they made the most important early studies of electric currents. They were able to do that because the electric battery was invented in 1799 by Alessandro Volta in Italy. Short bursts of electric currents were already known (Benjamin Franklin, for example, had demonstrated that lighting is an electric current) but one can't find out much from a short burst. With a battery, one can generate a continuous current whose effects can be investigated systematically.

Oersted and Faraday lived after what is called the Enlightenment, which had taken place earlier in the eighteenth century. This was a transition to modern ways of thinking, with less emphasis on religion and more on rational discussion.

7.2 OERSTED

Hans Christian Oersted was born in 1777 on the island of Langeland, which is in the North Sea and is part of Denmark. He died in Copenhagen, Denmark's capital, in 1851. Oersted's father was the village apothecary with a slender income and there was no school on the island. Hans and his slightly younger brother Anders were therefore educated on the island, by a wigmaker called Oldenburg and his wife. From Oldenburg they learned German, and from his wife they learned Danish reading and writing. The wife also taught them addition and subtraction, and a school friend taught them multiplication and division. The boys were further educated by the pastor, a surveyor and a university student, and any books they could get hold of. Whatever one brother learned he passed on to the other. Hans also learned some basic chemistry while helping in his father's business.

Hans Christian Oersted

In 1794 the brothers entered the University of Copenhagen. Hans started a degree in pharmacology, while his brother started one in law and was to eventually become the Prime Minister of Denmark. The brothers lodged together in a dormitory and dined at the home of their aunt. They received a small government grant which they supplemented by tutoring, but they still had to live frugally. Hans graduated with high honours in 1797 and obtained a PhD in philosophy in 1799.

In 1800 he became, for one year, the manager of the Lion Pharmacy in Copenhagen. The pharmacy was owned by Manthey, the Professor of Surgery in the University, and Oersted managed it because Manthey was travelling in Europe. Managing the pharmacy allowed Oersted to experiment with electric batteries. At the same time he gave some unpaid lectures at the University of Copenhagen.

In 1801 Oersted published a paper describing his investigation of batteries, and a method for measuring the current flowing through a wire. Also in 1801, Oersted himself received a travel scholarship and a government grant, which allowed him to travel in Europe for three years. In Germany he met the physicist Johann Ritter, who believed that there was a connection between electricity and magnetism. This made sense to Oersted from a philosophical viewpoint and it drew him towards physics. During the visit, Oersted helped Ritter by translating into French an essay that he had written describing a new type of battery. Ritter submitted the essay for a competition, but there is no record that it won.

On his way back from Paris Oersted visited Brussels, Leiden, Haarlem and Amsterdam. After returning from his travels, he was funded by the government to continue his research.

He sought a University Professorship, but didn't immediately get it. Instead he gave lectures privately, charging admission. The lectures became popular and in 1806 he became a Professor of Physics at Copenhagen University. He also gave lectures at the Military School. Oersted was an excellent lecturer and students flocked to his classes. Sometimes he lectured for as many as five hours a day. In addition to lecturing he established physics and chemistry laboratories for research and teaching. As a professor in Denmark's only university, he was one of the intellectual elite and he made several close friends.

In 1812 and 1813 he again travelled, to France and to Germany. At the same time he wrote a book expressing the view that electricity and magnetism were somehow related. In Berlin he published an article in German on chemistry, whose French version had already been published in Paris. On returning, Oersted took part in several public discussions. In one of them he expressed his opposition to dogma, and in another he likened the pursuit of science to an act of worship.

In 1814 Oersted married Inger Birgitte Ballum, the daughter of a pastor, and in the following years the couple had three sons and four daughters. In 1819 he was one of three scientists who, by royal command, went on a geological expedition to the island of Bornholm in the Baltic Sea.

In 1820, when he was forty-two years old, he made his great discovery. It was that a wire carrying an electric current is surrounded by a magnetic field. He had actually tried to detect such a magnetic field before, but without success because he had oriented the compass so that it would detect a field pointing away from the wire. The field actually points around a circle, whose plane is perpendicular to the wire, and he detected it only when he correctly oriented the compass. Oersted described this discovery in a four-page Latin document, which was widely circulated and made him famous.

Oersted continued scientific work after his great discovery, and in 1825 he had another first. Starting with aluminium chloride he was the first person to produce aluminium,.

Towards the end of his life Oersted wrote the following to a friend

> In my family I am as happy as a man can be. I have a wife
> whom I love, and children who are dear to me and prosper. I have

three sons–of whom on is of age and is employed in the forestry service of the King–and four daughters, of whom the eldest three are either married or betrothed. My brother, who for some time was a Commissioner for the King in our provincial parliament, has recently become a Minister of State. As for me, I am still a professor and director of the Polytechnic School, and Secretary of the [Danish] Royal Society of Sciences.

For more on Oersted, see [119]–[122].

7.3   FARADAY'S LIFE

Michael Faraday

Few scientists have come from a working-class background, fewer have been poorly educated, and practically none have found learning a difficult process. Michael Faraday confounded all three of these expectations.

He was born in 1791 in Newington Butts, Southwark, and he died in 1867 at Hampton Court, Surrey. When he was born Southwark was on the southern edge of London. Although London was the richest and most powerful city in the world, one could still walk across it in less than an hour from east to west and in a much shorter time from north to south.

The Britain that Faraday inhabited was very different from Newton's Britain. The practices of hanging, drawing and quartering, and of burning at the stake, had been replaced by simply hanging. By 1861 there were only four capital offences: Murder, High Treason, Piracy, and Arson in a Royal Dockyard. About half of the population were literate when Faraday was born, and about three-quarters by the time that he died. The telegraph was allowing instant long-distance communication. During Faraday's lifetime, the advent of the railway and of steamships facilitated long-distance travel.

Faraday's father James had been a blacksmith at Outhgill near Kirkby Stephen, in Westmorland which is now part of Cumbria. He had married Faraday's mother Margaret 1786, and they had moved to Newington Butts soon afterward. When Faraday was five they moved to Weymouth Street on the western edge of London, which was close to the premises of the ironmonger for which James worked. Faraday's father died in 1810 when Faraday was nineteen. His mother stayed in the house in Weymouth Street, taking in lodgers and dying there in 1838.

Faraday left school at the age of thirteen, barely able to read and write. He had a poor memory and found spelling and punctuation difficult as well as mathematics. On leaving school he worked for a bookseller and stationer name George Riebau and a year later he became an apprentice, to learn 'the Art of Bookbinding Stationary and Bookselling' as the indenture stated. Furthermore, 'in consideration of his faithful service no premium is given'–a significant saving for his father. The indenture also stated that Faraday had promised to serve Riebau faithfully, and to not commit fornication or haunt taverns or playhouses, etc.

Faraday moved to Blandford Street, where Riebau housed and fed him. During his seven-year apprenticeship Faraday was allowed to take home books and he became a good enough reader to benefit from them. His reading included books on science and he became interested in electricity. Riebau allowed him to conduct experiments in his shop, where there was some equipment including electric batteries.

In 1811, when Faraday was twenty, he attended lectures on science. They were given in his house by the silversmith John Tatum. Advertisements for them stated that they were open to 'both ladies and gentlemen' on the payment of 'one shilling per lecture'. Faraday's brother Robert paid for some of them. Faraday's four volumes of notes record his attendance at thirteen lectures, of which seven were on electrical subjects. (He had evidently improved his writing skills!) At the lectures and at the associated City Philosophical Society, Faraday made some life-long friends. Among them were Benjamin Abbott who later became a teacher, John Huxtable who lent Faraday some chemical textbooks and became an apothecary, Edward Magrath who later ran the Athenaeum, and Richard Phillips who became a distinguished chemist.

Also in 1811, Faraday's apprenticeship ended and he started work with a bookbinder called Henri De La Roche. At the same time he went to live with his widowed mother. De La Roche seems to have been difficult to work with, but he valued Faraday and even offered to leave him his business if he would stay with him. Faraday, though, had other ideas; he had decided to be a scientist.

Faraday's study of science was advanced in 1812, when Riebau showed Faraday's notes of Tatum's lectures to one of his customers. This particular customer was the son of William Dance, a violinist who lived nearby in Manchester Square. The following day, William Dance went to Riebau, who showed him the notebooks. Dance was so impressed that he gave Faraday tickets to attend the last four lectures. They were to be delivered by Humphrey Davy, Professor of Chemistry at the Royal Institution, of which Dance had been a proprietor since 1805.

The Royal Institution had been founded in 1799 for

> diffusing the Knowledge, and facilitating the general Introduction, of Useful Mechanical Inventions and Improvements; and for teaching, by Courses of Philosophical Lectures and Experiments, the application of Science to the common Purposes of Life.

So that lectures could be accompanied by spectacular experiments, the Royal Institution acquired the best laboratory in England, and one of the best in Europe. In it, Davy used the electric battery to isolate for the first time a number of chemical elements including sodium and potassium.

After attending the lectures, Faraday sent Davy a 300-page book based on notes that he had taken during these lectures. When Davy damaged his eyesight in an accident with nitrogen trichloride, those notes motivated him to take on Faraday as his assistant. Before long they were both injured in another explosion of nitrogen trichloride even though they were wearing masks. They wisely abandoned work on this dangerous substance.

Through Davy, Faraday was appointed in 1813 to the post of Chemical Assistant at the Royal Institution. When Davy set out on an eighteen-month tour of the continent in 1813–1815, his valet didn't want to go, so instead, Faraday went as Davy's scientific assistant and was asked to act as Davy's valet until a replacement could be found in Paris. None could be found so that Faraday was forced to fill the role of valet as well as assistant throughout the trip. Davy's wife, Jane Apreece, refused to treat Faraday as an equal making him travel outside the coach and eat with the servants. Faraday wasn't happy about that, but still the trip was worthwhile, because he met scientists in Europe and was exposed to some stimulating ideas.

After the trip Faraday continued to work at the Royal Institution, helping with experiments for Davy and other scientists. In 1821 he married Sarah Barnard. They met through their families at the Sandemanian church, which was an offshoot of the Church of Scotland. Until well after his marriage he served as deacon, which meant that he undertook pastoral duties such as visiting the sick. The Sandemanian meeting house was originally located in Paul's Alley in the Barbican but it moved in 1862 to Barnsbury Grove, Islington and that was where Faraday served the final two years of his second term as elder.

In 1821 Faraday made the first of his ground-breaking discoveries, but he didn't do much more research for the rest of the decade, because he was busy with other projects. It seems that Davy involved Faraday in these projects because they added to Davy's reputation, without worrying about the fact that they took up a lot of Faraday's time. One of the projects was the founding of Athenaeum Club, which was intended to bring together intellectuals. Faraday became its secretary and he sent out several hundred printed letters inviting eminent people to join. Then he acknowledged those who did, and sent a reminder to those who had not responded listing those who had already joined. Faraday was offered £100 annually to be secretary but he declined. Then he was elected a member with the first year's subscription waived.

Faraday managed to make more important discoveries in 1831, despite being still involved with the Athenaeum. He advised on the ventilation and lighting of the Athenaeum clubhouse, and supported the election of new members. That stopped in 1851, when he decided that he could no longer afford the membership fee. Later he complained that belonging to the Athenaeum had cost him nearly £200 for which he had just one dinner!

Another project involved the improvement of optical glass. A joint committee of the Board of Longitude and the Admiralty, of which Davy was a member, had been asked to look into this. It was known that glass of high quality was produced by the Bavarian optician Joseph Fraunhofer, but he kept his method secret. Faraday was asked to reproduce his method and in December 1827 he began a two-year investigation that took nearly two-thirds of his time. It ended in failure.

Things improved for Faraday when, in December 1829, he was appointed part-time Professor of Chemistry at the Royal Military Academy in Woolwich. From then until 1851, he spent two days a week at the Academy during their terms. That still took a lot of time, but it was a big improvement on the glass-making episode. For two decades cadets of the Royal Artillery and the Royal Engineers learned their chemistry from Faraday.

From the 1830s to the 1850s, Faraday was the go-to person for the government and its agencies whenever they wanted advice on scientific matters. In the 1830s he advised the Admiralty on whether oatmeal on prison transports was contaminated, on new methods of treating dry rot, and on the usefulness of lightning rods on ships. In the 1840s, he advised on possible naval uses of the electric telegraph. He also advised the East India Company on protecting their gunpowder stores from lightning, but by the end of that decade, he had had enough of the Admiralty, and declined to comment on the quality of disinfecting fluids.

One of Faraday's most important assignments for the Admiralty, was to advise about a proposed attack on the Russian island naval fortress of Cronstadt in the Baltic. In the spring of 1854 the UK and France declared war on Russia, fearing that it would gain from the decline of the Ottoman empire. That war's called the Crimean War, and it saw the largest deployment of British forces in Europe between 1815 and 1914. Part of Anglo-French strategy was to reduce Russian naval power, which led to a plan to attack Cronstadt. To further that, Dundonald (on whom the fictional Horatio Hornblower was based) proposed the use of sulphur-filled fireships to attack Cronstadt. The idea was that they would incapacitate or kill the defenders, allowing marines to land and capture the fortress.

An Admiralty committee was formed to consider this proposal, which sought Faraday's advice. He concluded that the proposal was 'correct in theory, but in its results must depend entirely on practical points'. He declared that these were 'untried and unknown'. As a result of the advice Cronstadt was never attacked, though Britain went on to win the war. A few months after Faraday's criticisms of Dundonald's proposal, a fire in Newcastle burned 2,000 tons of sulphur, and Faraday wrote to the Admiralty pointing out that it hadn't done much damage.

Faraday's advice here is typical of his approach. He was always cautious, with emphasis on the practicalities. And there's something else, which points to another side of Faraday's character. He was also asked to advise on the production of chemical weapons to be used in the Crimean War, but he declined on ethical grounds.

Faraday was occasionally asked to find out why something had happened, especially a disaster. The most important such occasion was when there was an inquiry into the explosion at Haswell Colliery, County Durham. It had happened on 28 September 1844, killing almost a hundred men and boys. The inquest was convened in Haswell on 30 September. The explosion had occurred during a period of intense industrial unrest in the Durham coalfield and the solicitor representing the relatives of those killed asked Prime Minister Robert Peel for government representatives to be sent to the inquest. Peel agreed and Faraday was appointed, along with Charles Lyell who was a geologist, and Samuel Stutchbury who was a mining engineer from Cornwall.

On 8 October, Faraday and Lyell travelled to Haswell and the inquest resumed on the next day. Lyell, who had originally trained as a lawyer, later recounted that Faraday cross-examined the witnesses with 'as much tact, skill, and self-possession as if he had been an old practitioner at the Bar'. On 10 October Faraday, Lyell, and Stutchbury investigated the mine's air flows and identified some laxity in the safety procedures. At one point Faraday found that he was sitting on a bag of gunpowder while a naked candle was in use: 'He sprung up on his feet, and, in a most animated and expressive style, expostulated with them for their carelessness'. The following day the jury returned verdicts of accidental death, which Faraday noted with the comment 'fully agree with them'.

After generously contributing to the subscription fund for the widows and orphans, Faraday and Lyell returned to London and submitted their report nine days later. Noting that coal dust had been a cause of the explosion they recommended that it should be drawn away from mines, but the mine owners reacted unfavourably because of the costs involved. Peel therefore tabled its consideration in the House of Commons on the same day as the second reading of a highly contentious bill to increase the grant to the Roman Catholic seminary at Maynooth. As a result of this strategy, the report received no more parliamentary attention, and the risk from coal dust was ignored until the Senghenydd Colliery Disaster of 1913.

From 1836 to 1885, Faraday was scientific adviser to Trinity House which is responsible for lighthouses. He developed a chimney to carry away the products of the oil lamps, which allowed his brother Robert to become a gas-lighting contractor, and which he patented. With two engineers from Chance Brothers, he manufactured the high-quality optical glass which was required by Chance for its lighthouses. Faraday also oversaw the first use of electricity to provide light in lighthouses using a carbon arc. An electric generator was constructed, utilising one of Faraday's scientific discoveries, which was installed in the South Foreland lighthouse and, in Faraday's presence, shone across the English Channel. Faraday visited the lighthouse regularly; on one occasion it was snowed up and he reached it 'by climbing over hedges, walls, and fields'. In the end though Faraday came out against the use of electric light, writing the following

> Much, therefore, as I desire to see the Electric light made available in lighthouses, I cannot recommend its adoption under present circumstances. There is no human arrangement that requires more regularity and certainty of service than a lighthouse. It is trusted by the Mariner as if it were a law of nature; and as the Sun sets so he expects that, with the same certainty, the lights will appear.

This method of lighting was subsequently abandoned, until the advent in the the 1920s of the light bulb and the central generation of electricity.

Faraday investigated other things too. One was the protection of the bottoms of ships from corrosion. His workshop still stands at Trinity Buoy Wharf, next to London's only lighthouse. He investigated industrial pollution at Swansea, and was consulted on air pollution at the Royal Mint. In July 1855 Faraday wrote a letter to the *Times* newspaper on the subject of the foul condition of the River Thames, which resulted in an often-reprinted cartoon in *Punch*.

He was appointed Assistant Superintendent of the House of the Royal Institution in 1821. He was elected a member of the Royal Society in 1824. (The Royal Society had been founded in 1660 by Charles II, and is the oldest scientific institution in the world.) In 1825 he became Director of the Laboratory of the Royal Institution. Six years later, in 1833, Faraday became the first Fullerian Professor of Chemistry at the Royal Institution, a position to which he was appointed for life without the obligation to deliver lectures. His sponsor and mentor was John 'Mad Jack' Fuller, a former Member of Parliament who created the position at the Royal Institution for Faraday.

Faraday was regarded as the outstanding scientific lecturer of his time. In 1826 he founded the Royal Institution's Friday Evening Discourses, and in the same year the Christmas Lectures for young people. They both continue to this day. He himself gave the Christmas lectures between 1827 and 1860. The twin objectives of the Christmas lectures were to present science to the general public in the hopes of inspiring them, and to generate revenue for the Royal Institution. They were notable events on the social calendar among London's gentry.

Over the course of several letters to his close friend Benjamin Abbott, Faraday outlined his recommendations on the art of lecturing: Faraday wrote 'a flame should be lighted at the commencement and kept alive with unremitting splendour to the end'. His lectures were joyful and juvenile. He filled soap bubbles with various gasses, to see if they were magnetic or not, and he marveled at the rich colours of polarised lights, but the lectures were also thought-provoking. In one of them, he said 'you know very well that ice floats upon water ... Why does the ice float? Think of that, and philosophise'. The lectures were called *The Rudiments of Chemistry, First Principles of Electricity, The Chemical History of a Candle, Attractive Forces, Voltaic Electricity, The Chemistry of Combustion, The Distinctive Properties of the Common Metals, Static Electricity, The Metallic Properties* and *The Various Forces of Matter and Their Relations to Each Other.*

Between 1836 and 1865 he was the Professor of Chemistry at the Royal Military Academy in Woolwich. Faraday assisted with the planning and judging of exhibits for the Great Exhibition of 1851 in London. He also advised the National Gallery on the cleaning and protection of its art collection, and served on the National Gallery Site Commission in 1857.

Education was another interest of Faraday's. He lectured on that in 1854 at the Royal Institution, and in 1862 he appeared before a Public Schools Commission to give his views on education in Great Britain. Faraday also spoke against the public's fascination with table-turning, mesmerism, and seances.

He was offered a knighthood in recognition for his services to science, but turned it down believing that it was against the word of the Bible to accumulate riches and pursue worldly reward. He stated that he wished to remain 'plain Mr Faraday to the end'. Elected a member of the Royal Society in 1824, he twice refused to become President.

Faraday suffered a nervous breakdown in 1839 but returned to his investigations into electromagnetism. In the early 1840s though, his health deteriorated and he began to do less research. In 1844 he was suspended as an elder of the Sandemanian church for missing a single Sunday service – the only one that he missed in his entire life. He explained that he had been dining with Queen Victoria, but the church elders were not impressed by this flimsy excuse!

In 1848 Faraday was awarded a grace and favour house in Hampton Court in Middlesex, free of all expenses and upkeep. In 1858 he retired to live there, dying on 25 August 1867 at the age of seventy-five. He had some years before declined an offer of burial in Westminster Abbey, but there is a memorial plaque there near Isaac Newton's tomb. Faraday was buried in the Dissenters' section of Highgate Cemetery.

7.4   FARADAY'S SCIENCE

It seems remarkable that Faraday made such profound discoveries, without any knowledge of mathematics. In a letter to André-Marie Ampère in September 1822, he explained his method

> I am unfortunate in a want of mathematical knowledge and the power of entering with facility any abstract reasoning. I am obliged to feel my way by facts closely placed together.

Faraday's scientific work went beyond his electromagnetic discoveries. He was the first person to liquify chlorine gas, the discoverer of benzene, the author of what are called oxidation numbers and the inventor of an early form of the Bunsen burner. I here focus on electromagnetism.

Early in 1821, Faraday was commissioned by his old City Philosophical Society friend Richard Philips to review for *Annals of Philosophy* the literature on electromagnetic phenomena. That was because of the announcement by Oersted of his discovery. Finding some of the descriptions of experiments unclear, Faraday repeated them. Among other things, he confirmed a discovery that had been made a decade earlier by the English scientist William Wollaston; that a current-carrying wire, suspended over a magnet into a bath of mercury, would move around the magnet. Faraday did not deny Wollaston's priority. It was Faraday, though, who used this phenomenon to build the first electric motor, which he demonstrated to the Royal Society in 1821.

After testing some new ideas that didn't work out, Faraday struck gold in 1831. On 28 August, he discovered the principle behind the electric transformers, which has allowed electricity to go from being a curiosity to a powerful new technology. This is recorded in his laboratory notebook, among a set of experiments with the heading: *Expts. On the production of Electricity from Magnetism, etc etc.* The apparatus, which he sketched in his notebook, comprised an iron ring wound with two coils of wire on opposite sides. When he switched on an electric current through one coil, he detected a burst of electric in the other coil, and when he switched it off he found another burst. (The existence of the current was established by using a galvanometer, which I'm going to describe in the chapter on Ampère.) He gave this effect the name 'induction' which remains the standard terminology. To be precise, induction is defined as the production of an electric current (or, more generally, of an electric field) by a magnetic field which is varying with time.

Three days later Faraday was in Hastings, on a holiday which lasted until 21 September. Not certain that he had found induction, he told Richard Phillips on the day after his fortieth birthday

> I have got hold of a good thing but can't say; it may be a weed instead of a fish that after all my labour I may at last pull up.

Returning on 21 September he made another discovery. He found that he could generated an electric current in a coil, by moving a magnet in and out of it. That is the principle behind electric generators. Two ground-breaking discoveries in less than four weeks!

On 28 October he made a third discovery. Working with the large array of magnets, he found that by rotating a copper disk in a magnetic field, he caused a current to flow in the disk. (That is of course, another example of induction. A point on the disk experiences a magnetic field which varies with time.)

He submitted these results to the Royal Society, in a paper called *Experimental Researches in Electricity*, and he then took a holiday in Brighton, remaining there until 1 December. The first part of the paper was read to the Royal Society on 14 November, and the other two parts on the 8 and 15 of December. Unfortunately, publication was delayed because the Royal Society had instituted a refereeing procedure.

Faraday now wrote to the Parisian savant J. N. P. Hachette describing his results. Hachette read the letter to the Académie des Sciences and it was reported in various Parisian newspapers. On New Year's Day, *Le Lycée* reported Faraday's discoveries, but claimed that the discovery had already been made in France. On 6 January 1832 the *London Morning Advertiser* published a translation of one of these reports.

Faraday was very concerned about this. He wrote on 14 January to the Secretary of the Royal Society urging him to try and speed up publication of his paper 'or else these philosophers may get some of my facts in conversation repeat them & publish in their own name before I am out'. His fears were justified, because others, such as Leopoldo Nobili in Florence, were given credit for Faraday's work. Faraday wrote to the *Literary Gazette* asserting his priority, and concluding with the following words

> I never took more pains to be quite independent of other persons than in the present investigation; and I have never been more annoyed about any paper than the present by the variety of circumstances which have arisen seeming to imply that I had been anticipated.

In 1832, a year after making all these discoveries, Faraday started to keep a record of his scientific thinking and investigations. Each paragraph was numbered and indexed, ending with number 16,041 when the record ended 28 years later.

In June 1845, Faraday was at the British Association's Annual Meeting in Cambridge. There he met William Thomson. Thomson, who was about to celebrate his twenty-first birthday, had just been elected a fellow of his college, Peterhouse. As an arrogant eighteen-year-old undergraduate, he had dismissed Faraday's approach to science, writing the following.

> I have been very much disgusted with his [Faraday's] way of *speaking* of the phenomena, for his theory can be called nothing else.

Now though they formed a close friendship, conducted mostly by correspondence following Thomson's appointment as Professor of Natural Philosophy at Glasgow University in 1846. Thomson would call on Faraday when he was in London, and at the 1859 meeting of the British Association in Aberdeen, he arranged for the Faradays to stay with his wife's uncle.

It was a question that Thomson put to him at the British Association meeting, that indirectly led Faraday to his final great discovery. He made it using what is called lead borate glass, which he had manufactured for use in lighthouses. What he found, was that if the glass is placed between the poles of a horseshoe magnet and polarised light is passed through it, its plane of polarisation is changed. Don't worry about the meanings of 'polarised light' and 'plane of polarisation', because they don't matter. What matters, is that light is in some way altered when it passes between the poles of the magnet. This was the first time that any connection between light and magnetism had been found, and it came as a great surprise.

For more information about Faraday's life and work, see [1], [2] and [123]–[126].

CHAPTER 8

Ampère

The experimental investigation by which Ampère established the law of the mechanical action between electric currents is one of the most brilliant achievements in science. The whole theory and experiment, seems as if it had leaped, full grown and full armed, from the brain of the Newton of Electricity. (Passage in A Treatise on Electricity and Magnetism *(1873) by James Clerk Maxwell.)*

8.1 BIOGRAPHY

André-Marie Ampère was born in Lyon on 20 January 1775. He died in Marseille on 10 June 1836 at the age of sixty-one. You invoke his name if you buy a 15 amp fuse, because the amp unit of electric current is named after him.

Ampère's father was Jean-Jacques Ampère, and his mother was Jeanne Antoinette Desutières–Sarcey Ampère. Both sides of the family were dealers in silk, and Ampère's father was a successful businessman

Ampère's mother was a devout woman, and he was initiated into the Catholic faith. He was to waver for a while during the terrible period of the French Revolution, but returned to his faith, entering a church on most days to pray with his rosary.

Ampère had an older sister Antoinette, and a sister Joséphine who was ten years younger. He spent the first seven years of his life in Lyon, except for the summers when the family moved to their property in Poleymieux-au-Mont-d'Or, a picturesque village a few miles up the river Saône from Lyon. When he was seven, Ampère's family took up permanent residence in Poleymieux, his father then spending only short periods in Lyon.

Ampère's father was an admirer of the eighteenth-century writer Jean-Jacques Rousseau, whose theories of education as outlined in *Émile* were the basis of Ampère's education. Rousseau believed that young boys should avoid formal schooling and pursue instead an 'education direct from nature'. Ampère's father therefore allowed his son to educate himself by means of his well-stocked library. Later on Ampère acquired other books.

André-Marie Ampère

His reading was wide in scope. He read various French Enlightenment authors, including Georges-Louis Leclerc and Compte de Buffon, who were primarily naturalists but also mathematicians and cosmologists. He read *Méchanique Analitique* by the Italian mathematician and astronomer Lagrange (born Lagrangia). He read the *Encyclopédie* founded by the art critic Denis Diderot and by the mathematician, mechanician, physicist, philosopher and music theorist Jean le Rond d'Alembert. The *Encyclopédie*'s view on equality was one of the inspirations for the French Revolution. He also taught himself Latin, so that he could read the works of two Swiss mathematicians and physicists, Daniel Bernoulli and Leonhard Euler. Euler is generally regarded as the greatest mathematician of all time.

Ampère was interested in mathematics even as a small child. According to one account

> The first marvellous faculty that began to develop in him was an uncontrollable tendency to arithmetical expression. Before he knew how to make figures, he had invented for himself a method of doing even rather complicated problems in arithmetic by the aid of a number of pebbles or peas. During an illness that overtook him as a child, his mother, anxious because of the possible evil effects upon his health of mental work, took his pebbles away from him.

Ampère began teaching himself mathematics at age thirteen. At about that time, his father took him to Lyon to receive lessons (including some on calculus) from Abbot Daburan who was a professor of theology at the Collège de la Trinité. Then he took him to the Collège de Lyon to attend some physics lectures.

When Ampère was fourteen the French Revolution began. His father was called into public service by the new revolutionary government, becoming a justice of the peace in a small town near Lyon. When the Jacobin faction seized control of the Revolutionary government in 1792, his father resisted the new political tides, and was guillotined on 24 November 1793. The night before his death, his father wrote to his mother 'I desire my death to be the seal of a general reconciliation between all our brothers; I pardon those who rejoice in it, those who provoked it, and those who ordered it....' Fortunately, the family property had been transferred to his mother's name, which meant that they were able to keep it.

Ampère was then eighteen and, devastated by the death of his father, he discontinued his studies for a year. He recalled later that two things helped to lift him from depression: a renewed interest in botany and *Corpus poetarum latinorum* which contained the work of Latin poets.

In 1796 Ampère met Julie Carron, the orphaned daughter of a silk merchant. In 1797 he advertised himself in Lyon as a private tutor in mathematics. This enterprise was very successful, and brought him to the attention of Lyon's intellectuals. He took his first regular job in 1799 as a mathematics teacher, which gave him the financial security to marry Carron and father his first child, Jean-Jacques.

This was the time of the transition to the Napoleonic regime in France, which gave Ampère new opportunities. In 1802, despite having no formal training, he became a professor of physics at the École Centrale in Bourg-en-Bresse, 60 kilometres from Lyon, which had been staffed by Jesuits before the Revolution.

When he moved, Ampère left his wife and infant son in Lyon. That may have been because his wife was ill, and indeed she died in July the following year. On the day of her death he wrote two verses from the Psalms, and the prayer, 'O Lord, God of Mercy, unite me in Heaven with those whom you have permitted me to love on earth'.

In 1804 Ampère joined the University of Paris as a répétiteure in analysis. Analysis is a branch of mathematics, and his position meant that he was tutor to students who received lectures from the professor of analysis. The professor, when he arrived, was the famous Augustin-Louis Cauchy. Ampère was to remain in Paris until he died at the age of sixty-one. During that time, he would have witnessed Napoleon's coronation as Emperor, his military success and then defeat, the coronation of Louis XVIII, Napoleon's brief return, the restoration of the monarchy in 1815 and the revolution of 1830.

On 1 August 1806 Ampère married Jennie Potot. His biographer Launay viewed the match as unsuitable, writing

> It was a household as bourgeois as possible, in the pejorative sense
> in which artists understand that word: narrow ideas, prejudices,
> pretensions, living only for money and vanity, not having the least
> ideas about the sciences, exactly the opposite of what it should
> have been for a big child as simple and modest as Ampère.

They had a daughter in July 1807, but by that time they had separated at his wife's insistence. During the autumn of 1807, Ampère was joined in Paris by his mother, his sister Joséphine and his son Jean-Jacques who was seven. Ampère sent his son to boarding school, and even during the summer holidays he didn't see much of him, because he was travelling in his capacity as a University Inspector.

In July 1808 Ampère got a separation decree from Jennie Potot, which gave him custody of their daughter. His mother died in the following year. In the next few years Ampère had relationships with a number of women. One woman, whose name we do not know, was particularly important to him, but she entered in 1811 into a marriage of convenience with an elderly but wealthy man from the provinces.

In 1809, again notwithstanding his lack of formal qualifications, Ampère was appointed a professor of mathematics at the École Polytechnique. As the position did not bring in sufficient income, Ampère also took on the position of inspector-general for the newly formed university system. In 1814 he was invited to join the newly formed *Institute Impérial*, the umbrella under which the reformed state Academy of Sciences would sit. In 1819 and 1820 he offered courses in respectively philosophy and astronomy, at the University of Paris.

In 1824 he was elected to the prestigious chair in experimental physics at the Collège de France. Ampère's teaching duties took up a lot of his time, but he still managed to engage in research.

During the last decade of his life, Ampère lived with his sister and his son, but the arrangement didn't work very well. The sister ran up large debts in maintaining their household, while the son Jean-Jacques used his inheritance from Ampère's mother to enjoy leisurely trips abroad. Ampère's relationship with the son had always been stormy, because both men were temperamental, subject to long periods of brooding followed by explosions of anger.

The following episode, though, shows that Ampère really cared for his son. In 1820, Jean-Jacques met the famous Madame Récamier, whose portrait hangs in the Louvre. Her salon drew the leading literary and political figures of Paris, including the writer Balanche and, sporadically, the writer, politician, diplomat and historian Chateaubriand. Nourished by this literary circle, and encouraged by his father, Jean-Jacques completed in 1823 his first drama, *Rosemonde*, and his father began organising its staging. Unfortunately, his efforts were wasted because Madame Récamier went to Italy for a year and Jean-Jacques followed her. Eventually though, Jean-Jacques became successful as a literary historian.

Ampère's daughter Albine also caused him concern. In 1827, she had married Ride, a violent alcoholic who was mentally unstable. Beginning in early 1830, he was repeatedly placed in sanitoriums for treatment of his alcoholism but he always relapsed. During October 1930 Albine fled to her father's house, but Ride arrived too. After that the couple at times lived with Ampère. His account of the ending of that arrangement is described in a letter to Jean-Jacques, who had been living with Ampère until the arrival of Ride, in the following words

> All my life I will remember the danger I ran shortly before my departure from Paris, when he came into my room one night with his sword – a danger from which I escaped only by indulging his mania, appearing to agree with his ideas and persuading him to have the porter come to our defense. He then opened the door to the street where he ran in his nightshirt with his bare sword ever in his hand.

Police disarmed and arrested the swordsman who joined his two brothers in New Orleans.

Throughout his life Ampère read widely, on topics ranging from history and travel, to poetry and philosophy. He had a remarkable memory, and could recite whole passages from the Encyclopédie, including those on obscure subjects such as heraldry and falconry. On the other hand, he was notoriously absent-minded. He is reported to have written some equations on the back of a carriage, only to see them whisked away. Another example was given by Camille Flammarian, who wrote

> ... Ampère, who one day, as he was going to his course of lectures, noticed a little pebble on the road; he picked it up, and examined with admiration the mottled veins. All at once the lecture which he ought to be attending to returned to his mind; he drew out his watch; perceiving that the hour approached, he hastily doubled his pace, carefully placed the pebble in his pocket, and threw his watch over the parapet of the Pont des Arts.

In 1836, while on an inspection tour in Marseilles, Ampère contracted a fever and died. He was buried in a simple fashion being unknown to the general public. In 1869, by which time I suppose his achievements were beginning to be recognised, his remains were transferred to the Montmartre Cemetery in Paris. In 1881 the Paris Congress of Electricians defined the ampere unit of electric current.

8.2   SCIENCE

In Bourg Ampère did research in mathematics, producing in 1802 *Considérations sur la théorie mathématique de jeu* (*Considerations on the Mathematical Theory of Games*) which he sent to the Paris Academy of Sciences in 1803. This was one of the first works on probability theory.

In September 1820 his friend (and eventual eulogist) François Arago showed the members of the French Academy of Sciences Oersted's surprising discovery, that a current-carrying wire generates a magnetic field. Ampère checked this using a compass needle. To eliminate the effect of the earth's magnetic field, the compass was placed so that the needle rotated in a plane that was perpendicular to the direction of that field. Ampère called his device a galvanometer after the eighteenth-century Italian researcher Luigi Galvani.

After further investigation Ampère found that a magnetic field exerts a force on a current-carrying wire. Combined with Oersted's discovery this means that two wires exert a force upon each other. Which raises the following question; just what force does one short piece of a current-carrying wire exert, upon another short piece of a current-carrying wire? The answer will obviously be quite complicated, because the force will depend on the directions of the short pieces, as well as on the distance between them. In October 1820 Ampère made a guess about the answer and then performed experiments to verify that it was correct. It was this discovery, made at the age of forty-four, that made him famous.

Later Ampère conjectured the existence of a charged particle, whose motion through the wire corresponds to an electric current. He called his particle the 'electrodynamic molecule' and we now call it the electron.

Ampère published his findings in 1826. This, his magnum opus, was entitled *Mémoire sur la théorie mathématique des phénomènes électrodynamiques uniquement déduite de l'experience (Memoir on the Mathematical Theory of Electrodynamic Phenomena, Uniquely*

Deduced from Experience). The work coined the term *electrodynamics* and provided the foundation for future studies of electricity and magnetism.

In 1832 there was an outbreak of cholera in Paris, and Ampère went to Claremont to escape it. There, his host Gonod helped him to compile the first volume of *Essai sur la Philsophie des Sciences* which reflects on his experiences as a scientist and suggests a classification of academic disciplines. It was published in 1834. Later Ampère wrote a second volume. That was compiled by his son, and published posthumously in 1843. Unfortunately, Ampère's classification has not been regarded as useful by the community so that his efforts in that direction were wasted.

For more information see the following; account of Ampère's early interest in mathematics and also the statement about his prayer with a rosary [127], Ampère's life and work [128, 129], the book *Corpus poetarum latinorum* [130], *Émile ou de l'éducation* by Jean-Jacques Rousseau [131, 132, 133, 134], Lagrange's *Méchanique Analitique* [135, 136, 137], the *Encyclopédie* [138], Flammarion's astronomy book mentioning Ampère's absent-mindedness [139], Antoine Leonard Thomas's *Eulogy of Descartes* [140] .

CHAPTER 9

Maxwell

I have also a paper afloat, with an electromagnetic theory of light, which, till I am convinced to the contrary, I hold to be great guns. (In a letter from Maxwell to C. H. Cay, written on 5 January 1865, as quoted in American Journal of Physics, *44(8), page 470.)*

9.1 BIOGRAPHY

James Clerk Maxwell was born in Edinburgh on 13 June 1831 and he died in Cambridge on 5 November 1879. His father was John Clerk Maxwell and his mother was Frances Cay.

John Clerk Maxwell was an advocate (defence lawyer). He was a member of the Clerk family of Penicuik that held the baronetcy of Clerk of Penicuik and the sixth Baronet was his uncle. He had been born 'John Clerk', the surname Maxwell having been added when he inherited (as an infant in 1793) the

James Clerk Maxwell

Middlebie country estate near Corsock. Corsock is in Kirkcudbrightshire, in the Galloway region of South-West Scotland. Maxwell's mother came from Northumberland.

Maxwell's parents had met and married when they were well into their thirties, and his mother was nearly forty when he was born. There had been one earlier child who died in infancy. When Maxwell was young, his family moved to Glenlair House which his parents had built on the Middlebie estate. Being in charge of that estate was a significant responsibility, because about a thousand people lived in it. Maxwell played with the local children, acquiring a thick Galloway accent which he never entirely lost.

In Maxwell's day there were two Protestant Churches in Scotland. One was the Scottish Episcopalian Church, an extension of the Church of England which had been established by Henry VIII in 1534. The other was the Church of Scotland which was Presbyterian. As a child he attended services of both the Scottish Episcopalian Church (his father's denomination) and the Church of Scotland (his mother's denomination). Maxwell became a Presbyterian and underwent an evangelical conversion in April 1853. In his later years he became an Elder of the Church of Scotland.

Maxwell had an unquenchable curiosity from an early age. In a passage added to a letter from Maxwell's father to his sister-in-law Jane Cay in 1834, Maxwell's mother described his inquisitiveness at the age of three:

> He is a very happy man, and has improved much since the weather got moderate; he has great work with doors, locks, keys, etc., and "show me how it doos" is never out of his mouth. He also investigates the hidden course of streams and bell-wires, the way the water gets from the pond through the wall....

Recognising the potential of the young boy, Maxwell's mother Frances took responsibility for James' early education (which was in any case, at that time, usually the job of the woman of the house).

By the time he was eight, Maxwell could recite long passages of Milton and all 176 verses of the 119th psalm. He could also give chapter and verse for almost any quotation from the psalms.

At least while his mother lived, Maxwell's home was a happy place with music and dancing a common occurrence. Unfortunately, she was taken ill with abdominal cancer and, after an unsuccessful operation, died when Maxwell was eight. His education was then overseen by his father John and his mother's sister Jane. His schooling began unsuccessfully under the guidance of a sixteen-year-old hired tutor. The tutor treated Maxwell harshly and John dismissed him in November 1841. John then sent Maxwell to the prestigious Edinburgh Academy. In Edinburgh, Maxwell lived with his aunt Isabella. There, his passion for drawing was encouraged by his older cousin Jemima.

The ten-year-old Maxwell, having been raised in isolation on his father's countryside estate, didn't fit in well at the Edinburgh Academy. The first year had been full, obliging him to join the second year with classmates a year his senior. His mannerisms and Galloway accent struck the Edinburgh boys as rustic. Having arrived on his first day of school wearing a pair of homemade shoes and a tunic, he earned the unkind nickname of 'Daftie' and he often sat alone in a corner of the playground. After a couple of years of torment though, Maxwell went berserk and earned the grudging respect of his tormenters. He also made some friends including Lewis Campbell and Peter Guthrie Tait, two boys of a similar age who were to become notable scholars later in life. They remained lifelong friends thereafter.

Tait later described Maxwell's first days at school in the following words.

> At school he was at first regarded as shy and rather dull. He made
> no friendships and spent his occasional holidays in reading old
> ballads, drawing curious diagrams and making rude mechanical
> models. This absorption in such pursuits, totally unintelligible
> to his school fellows, who were totally ignorant of mathematics,
> procured him a not very complimentary nickname.

Maxwell won the school's scripture biography prize in his second year. At the age of thirteen he won the school's mathematical medal, and first prize for both English and poetry. He was athletic, being an accomplished horseman and fond of swimming and pole-vaulting.

He left the Academy in 1847 at age sixteen to attend classes at the University of Edinburgh. On leaving the Academy he wrote

> Dear old Academy,
> Queer old Academy,
> A merry lot we were, I wot,
> When at the old Academy
> ....
> Let pedants seek for scraps of Greek,
> Their lingo to Macademize;
> Gie me the sense, without pretence,
> That comes o' Scots Academies.
>
> Let scholars all, both grit and small,
> Of learning mourn the sad demise;
> That's as they think, but we will drink
> Good luck to Scots Academies.

According to the records of the University of Edinburgh, Maxwell borrowed the following books while as student: *Calcul Differential (Differential Calculus)* by Cauchy, *Geometrie Descriptive (Descriptive Geometry)* by Monge, *Mechanics* by Poisson, *Theorie de Chaleur (Theory of Heat)* by Fourier, *Optics* by Newton, *Scientific Memoirs* (edited and published by Taylor) and *Principles of Mechanism* by Willis. All of these authors were leaders in their field.

Maxwell's friends Campbell and Tait spent only a year at Edinburgh University before going to Oxford or Cambridge. Maxwell chose to spend the full three years there. That gave him the option of following his father into the law which was possible only if one had studied in a Scottish university.

In any case Edinburgh had been, since the second half of the previous century, one of the leading cultural cities of Europe. Famous thinkers such as David Hume and Adam Smith had met together on a regular basis and the Royal Society of Edinburgh had been set up in 1783. An English visitor, the poet Edward Topham, had written in 1775 the following

> The degree of attachment which is shewn to Music in general in this country, exceeds belief. It is not only the principal entertainment, but the constant topic of conversation; and it is necessary not only to be a lover of it, but to be possessed of a knowledge of the science to make yourself agreeable to society.

Because of all that Edinburgh was one of several European cities that was sometimes called the Athens of the North.

In October 1850, already an accomplished mathematician at the age of nineteen, Maxwell left Scotland for the University of Cambridge which offered at that time the best mathematical education in the English-speaking world. This was just before it became possible to travel from Edinburgh to London by train. Maxwell initially attended Peterhouse but before the end of his first term transferred to Trinity where he believed it would be easier to obtain a fellowship.

It was in Cambridge that Maxwell blossomed. According to one account 'Among his friends he was the most genial and amusing of companions, the propounder of many a strange theory'. At Cambridge he was tutored by the mathematician and geologist William Hopkins who was famously successful in preparing students for the final examinations (the tripos). A fellow member of Hopkins' team, W. N. Lawson, wrote the following in his diary, in 1853

He (Hopkins) was talking to me this evening about Maxwell. He says he is unquestionably the most extraordinary man he has met .... looks upon him as a great genius .... that one day will shine as a light in physical science.

Maxwell was in the habit of working late into the night at Cambridge, and he continued to do that for the rest of his working life. To enable it he slept between 5.00pm and 9.30pm. He pursued his Christian faith as well as science. He joined the 'Apostles', an exclusive debating society of the intellectual elite, where through his essays he sought to work out this understanding. He wrote the following

> Now my great plan, which was conceived of old, ... is to let nothing be wilfully left unexamined. Nothing is to be holy ground consecrated to Stationary Faith, whether positive or negative. All fallow land is to be ploughed up and a regular system of rotation followed. ... Never hide anything, be it weed or no, nor seem to wish it hidden. ... Again I assert the Right of Trespass on any plot of Holy Ground which any man has set apart. ... Now I am convinced that no one but a Christian can actually purge his land of these holy spots. ... I do not say that no Christians have enclosed places of this sort. Many have a great deal, and every one has some. But there are extensive and important tracts in the territory of the Scoffer, the Pantheist, the Quietist, Formalist, Dogmatist, Sensualist, and the rest, which are openly and solemnly Tabooed. ...
>
> Christianity — that is, the religion of the Bible -- is the only scheme or form of belief which disavows any possessions on such a tenure. Here alone all is free. You may fly to the ends of the world and find no God but the Author of Salvation. You may search the Scriptures and not find a text to stop you in your explorations. ...
>
> The Old Testament and the Mosaic Law and Judaism are commonly supposed to be "Tabooed" by the orthodox. Sceptics pretend to have read them, and have found certain witty objections ... which too many of the orthodox unread admit, and shut up the subject as haunted. But a Candle is coming to drive out all Ghosts and Bugbears. Let us follow the light.

In the summer of his third year Maxwell spent some time at the Suffolk home of the Rev C. B. Taylor, the uncle of a classmate, G. W. H. Taylor.

The love of God shown by the family apparently impressed Maxwell, particularly after he was nursed back from ill health by the minister and his wife.

On his return to Cambridge Maxwell wrote to his recent host a chatty and affectionate letter including the following testimony

> ... I have the capacity of being more wicked than any example that man could set me, and ... if I escape, it is only by God's grace helping me to get rid of myself, partially in science, more completely in society, — but not perfectly except by committing myself to God ...

In the summer of 1854, Maxwell and his fourteen-year-old cousin Elizabeth Cay fell in love but the relationship didn't continue. Later in 1854, Maxwell graduated from Trinity with a degree in mathematics. He scored second highest in the final examination, coming behind Edward Routh who himself became a noted tutor of pupils for the tripos. He was later declared equal with Routh in the more exacting Smith's Prize examination. Maxwell decided to remain at Trinity after graduating and applied for a fellowship, which was a process that he could expect to take a couple of years. This would leave him free, apart from some tutoring and examining duties, to pursue scientific interests at his own leisure.

In February 1855 Maxwell nursed his father through bronchitis and pneumonia. Maxwell was made a fellow of Trinity on 10 October 1855. In November he again became a nursemaid, this time to a friend who was ill with a fever. He also began to teach working men's evening classes and he undertook to write a textbook on optics which never saw the light of day. On becoming a fellow of Trinity he prepared lectures on hydrostatics and optics, and set examination papers.

The following February he was urged by Forbes to apply for the newly vacant Chair of Natural Philosophy at Marischal College, Aberdeen. His father assisted him in the task of preparing the necessary references but died on 2 April at Glenlair before either of them knew the result of Maxwell's candidacy. In the summer of 1856 Maxwell took a trip to Belfast to visit his cousin William Cay who was studying engineering there.

In 1856 Maxwell did became a Professor of Physics at Marischal College. In February 1857 he wrote to his Aunt Jane

I had a glorious solitary walk today in Kincardinshire by the coast – black cliffs and white breakers. I took my second dip this season. I have found a splendid place, sheltered and safe, with gymnastics on a pole afterwards.

On the other hand he also wrote:

No jokes of any kind are understood here. I have not made one for two months, and if I feel one coming I shall bite my tongue.

In 1857 he wrote an article on the theory of the gyroscope, to illustrate which he had a special top made in Aberdeen. He took that with him to Cambridge when he came up for his MA. He exhibited it at a tea–party in his room, and in the evening as his friends were leaving he set it spinning. Early next morning, seeing from his window one of them crossing the court towards his rooms, he quickly started up the top and jumped back into bed. He admitted the deception later but still his top became a legend.

At Marischal the twenty-five-year-old Maxwell was at least fifteen years younger than any other professor. Taking seriously his responsibilities as head of a department, he devised the syllabus and prepared lectures. He lectured for fifteen hours a week, including a weekly pro bono lecture to the local working men's college. He lived in Aberdeen during the six months of the academic year (November to April) and spent the rest of the time at Glenlair, which he had inherited from his father.

In 1857 Maxwell befriended the Reverend Daniel Dewar, who was then the Principal of Marischal. Through him Maxwell met Dewar's daughter, Katherine Mary Dewar. They were engaged in February 1858 and married in Aberdeen on 2 June 1858. Seven years Maxwell's senior, comparatively little is known of Katherine, although it is known that she helped in his lab and worked on experiments in viscosity. The couple had no children.

In 1859 the first cable message was transmitted across the Atlantic. A few months later the cable broke, which prompted Maxwell to write

Under the sea
No little signals are coming to me
Under the sea
Something has surely gone wrong.
And it's broke, broke, broke;
What is the cause of it does not transpire,

But something has broken the telegraph wire.

With a stroke, stroke, stroke.

Or else they've been pulling too strong.

In 1860 Marischal College merged with the neighbouring King's College to form the University of Aberdeen. There was no room for two professors of Natural Philosophy so Maxwell, despite his scientific reputation, found himself laid off. He was also unsuccessful in applying for Forbes's recently vacated chair at Edinburgh which went instead to a friend and colleague, Peter Guthrie Tait.

It seems that although Maxwell's eminence as a physicist was acknowledged, he was also known to be a poor teacher, whereas Tait was known to be an excellent one. This tipped the balance in Tait's favour. Indeed Maxwell would, according to a student Ambrose Fleming in the late 1890s, often be lecturing to two or three students. As Tait put it in an article in *Nature*

> The rapidity of this thinking, which he could not control, was such as to destroy, except for the highest class of students, the value of his lectures.

His elocution was also poor and he got himself into trouble on the blackboard. As we saw earlier, Newton was similarly unsuccessful as a lecturer.

After failing to obtain the Edinburgh post, Maxwell successfully applied for the Chair of Natural Philosophy at King's College, London. After recovering from a near-fatal bout of smallpox in 1860, Maxwell moved to London with his wife. Maxwell would often attend lectures at the Royal Institution, where he came into regular contact with Michael Faraday. Faraday was forty years Maxwell's senior and showed signs of senility, but the two nevertheless maintained a strong respect for each other's talents. One day, after attending a lecture at the Royal Institution, the two of them found themselves trapped in the slow-moving crowd, some distance apart. Faraday, mindful of Maxwell's work on the behaviour of molecules in a gas called out: 'Ho, Maxwell, can't you get out? If any man can find his way through a crowd, it should be you!'.

While the Maxwells were living in London, Katherine Maxwell's brother Donald Dewar came to London, for an operation at the hands of a well-known surgeon. While he was convalescing, Maxwell gave up the use of the ground floor of his house to him and personally helped to nurse him. It's said that the patient would await Maxwell's return from the college with an almost childlike expectation.

In 1865, Maxwell resigned the chair at King's College and returned to Glenlair with Katherine, working there for the next six years. Then, in 1871, he became the first Cavendish Professor of Experimental Physics at Cambridge (now called simply Cavendish Professor of Physics). Subsequent holders of the position have all been renowned physicists. John Strutt (Lord Raleigh) discovered the atmospheric element argon, as well as Rayleigh scattering which is the reason why the sky is blue. J. J. Thomson discovered the electron. Ernest Rutherford founded the science of nuclear physics. William Bragg founded, with his father, the science of X-ray crystallography. Nevill Mott did fundamental work on semi-conductors. Brian Pippard did fundamental work on superconductors. Sam Edwards did fundamental work in theoretical physics and in polymer science. The present holder Richard Friend has done fundamental work on many aspects of condensed matter physics (the physics of solids).

Maxwell was required to set up the Cavendish Laboratory, named (like the title of the professor) in honour of Henry Cavendish who was a notable scientist of the late eighteenth century. He closely supervised both the building of the laboratory and the purchase of the apparatus. One of Maxwell's last contributions to science was the editing (with copious original notes) of the research of Henry Cavendish, though in the opinion of some that was unnecessary because Cavendish's writing was already quite clear.

Maxwell died in Cambridge of abdominal cancer on 5 November 1879 at the age of forty-eight. His mother had died at the same age of the same type of cancer. The minister who regularly visited him in his last weeks was astonished at his lucidity and the immense power and scope of his memory, but comments more particularly,

> ... his illness drew out the whole heart and soul and spirit of the man: his firm and undoubting faith in the Incarnation and all its results; in the full sufficiency of the Atonement; in the work of the Holy Spirit. He had gauged and fathomed all the schemes and systems of philosophy, and had found them utterly empty and unsatisfying-'unworkable' was his own word about them — and he turned with simple faith to the Gospel of the Saviour.

As death approached Maxwell said the following to a Cambridge colleague

I have been thinking how very gently I have always been dealt

with. I have never had a violent shove all my life. The only desire
which I can have is like David to serve my own generation by the
will of God, and then fall asleep.

As a great lover of Scottish poetry, Maxwell memorised poems and
wrote his own. The best known is *Rigid Body Songs*, closely based on
Comin' Through the Rye by Robert Burns, which he apparently used
to sing while accompanying himself on a guitar. It has the following
opening lines, which exploit the fact that physicists call an object a
body if they are studying its movement.

> Gin a body meet a body
> Flyin' through the air.
> Gin a body hit a body,
> Will it fly? And where?

Maxwell is buried at Parton Kirk, close to where he grew up. A biography *The Life of James Clerk Maxwell*, by his former schoolfellow
and lifelong friend Professor Lewis Campbell, was published in 1882.
His collected works were issued in two volumes by the Cambridge University Press in 1890.

9.2 SCIENCE

When he was thirteen Maxwell rediscovered the shapes known as regular polyhedra. At the age of forteen he wrote *Oval Curves*. This paper
describes the method of drawing an ellipse, that I explained in the chapter on Kepler, and then describes more complicated constructions. It
was presented to the Royal Society of Edinburgh by the Scottish physicist and glaciologist James Forbes who was professor of natural philosophy at Edinburgh University. This was because Maxwell was deemed
too young to present the work himself.

After that, Maxwell was regularly taken by his father to meetings of
the Royal Society of Edinburgh where he heard about cutting-edge scientific research. When he was fifteen he heard a talk explaining how the
existence of the recently observed planet Neptune had been predicted
by analysing irregularities in the motion of the planet Uranus.

While he was at Edinburgh University, Maxwell received tuition
from some highly regarded people. One of them was James Forbes.
Another was the Scottish philosopher Sir William Hamilton. A third
was the English mathematician and theoretical physicist Philip Kelland. Even so Maxwell didn't find his classes very demanding. As a

result he was able to immerse himself in private study, during free time at the university and particularly when back home at Glenlair. At Glenlair he would experiment with improvised chemical, electric, and magnetic apparatus. In particular, he investigated the properties of polarised light, discovering what is called photoelasticity.

At age eighteen Maxwell contributed two papers to *Transactions of the Royal Society of Edinburgh*. Just as with the earlier paper he was again considered too young to stand at the rostrum himself. The papers were instead delivered by his tutor Kelland.

Immediately after earning his degree Maxwell read his paper *On the Transformation of Surfaces by Bending* to the Cambridge Philosophical Society. This is one of his few purely mathematical papers.

In March 1855 Maxwell wrote a paper *Experiments on Colour*, which laid out the principles of colour combination, and which he presented to the Royal Society of Edinburgh. In the same month a Cambridge board of examiners announced that a prize (namely, the fourth Adams Prize) would be given for an explanation of Saturn's rings. Between 1856 and 1858 Maxwell spent much of his time on this problem. The result, published in 1859 as *On the Stability and Motion of Saturn's Rings*, is a monumental ninety–page work which shows that the rings must be composed of a large number of separate solid objects. (Alternatives that people had considered, were that they were a continuous solid or a liquid.) He received the prize for this work which George Airy, an eminent contemporary, called 'one of the most remarkable applications of Mathematical Physics that I have ever seen'.

After working on Saturn's rings Maxwell turned his attention to the kinetic theory of gases. By 'kinetic theory' one means an understanding of the properties of a gas in terms of the movement of its constituent atoms or molecules. Maxwell considered the problem of finding the probability distribution of the speeds of the atoms or molecules. On the basis partly of guesswork, he arrived at a result which is called the Maxwell-Boltzmann distribution. (The name Boltzmann appears because it was he who proved that Maxwell was right.) Maxwell presented this in 1859, in a landmark paper *Illustrations of the Dynamical Theory of Gases*. He also performed experiments on friction within a gas.

Following in the footsteps of Isaac Newton, and of Thomas Young in the early 1800s, Maxwell also studied colour vision. His understanding of that subject allowed him to produce the world's first colour photograph, with the assistance of Thomas Sutton who invented the single-lens reflex camera. Maxwell showed this photograph in 1861 during a Royal Institution lecture.

Maxwell worked on other problems too. In 1868 he published *On Governors*. 'Governor' here refers to the centrifugal governor used to regulate steam engines. This paper was ignored at the time because it was difficult to understand but it's now regarded as the founding paper for what's called control theory. In 1870 he published *On reciprocal figures, frames and diagrams of forces* which discusses the rigidity of rod-and-joint frameworks like those in many bridges.

In 1871 he was the first to make explicit use of the technique in physics called dimensional analysis. Also in that year, he returned to the study of gases but now from the perspective of thermodynamics. Thermodynamics just studies the overall properties of a gas without worrying about the atoms of molecules. To be precise, it studies connections between the temperature, pressure, volume and entropy of the gas. Maxwell established what are called Maxwell's thermodynamic relations, involving those four quantities.

Finally, we come to Maxwell's greatest achievement, which was in the study of electromagnetism. He drew together everything that was known about the subject, in a set of equations that are mathematically equivalent to the set that we now call Maxwell's equations. The work was mostly done at Glenlair, during the period between holding his London post and his taking up the Cavendish chair. Having written down the equations there came to Maxwell a great surprise. He found that the equations predicted the existence of waves, consisting of oscillating electric and magnetic fields. Maxwell calculated the speed of his waves and found that it was the same as the speed of light. He declared that this could not be a coincidence

> ... it seems we have strong reason to conclude that light itself (including radiant heat, and other radiations if any) is an electromagnetic disturbance in the form of waves propagated through the electromagnetic field according to electromagnetic laws. (Passage in *A Dynamical Theory of the Electromagnetic Field, 1864.*)

So we finally knew what was light, a source of wonder through the ages!

As one sees from the quotation above, Maxwell recognised that 'radiant heat' is also an electromagnetic wave. By radiant heat, he meant what we now call infrared radiation, whose wavelength is just a bit longer than that of red light. By 'other radiations' he meant electromagnetic waves with wavelengths different from those of visible light and infrared radiation. Regarding his 'if any' we now know of electromagnetic radiation over a wide range of wavelengths. Going up in wavelength from infrared we come to microwaves and then to radio waves. Going down in wavelength from blue we come to ultraviolet, then X-rays, then gamma rays. All of these are generated by ourselves for our use and all of them are observed by astronomers, coming from a variety of sources.

For *A dynamical theory of the electromagnetic field* see [141]. For Maxwell's scientific papers see [142]. For accounts of his life and work see [143]–[147].

CHAPTER 10

Einstein

There is not the slightest indication that [nuclear energy] will ever be obtainable. It would mean that the atom would have to be shattered at will. (Einstein, as quoted in the Pittsburgh Post-Gazette, *29 December 1934.)*

10.1 BIOGRAPHY

Albert Einstein was born during the year of Maxwell's death, on 14 March 1879. He died on 18 April 1955 at the age of seventy-six. His lifetime saw the advent of the telephone, sound and visual recording, radio, television, and travel by car and aeroplane. It also saw, as a direct result of his work, the advent of nuclear power and nuclear bombs.

Albert Einstein

He was born in Ulm, which was part of the Kingdom of Württemberg that since 1871 had been part of the German Empire. (The German Kingdoms were finally abolished in 1918 and for a long time the kings had had very limited powers.) Ulm was in south-western Germany, in a region known as Swabia. His parents were Hermann Einstein, a salesman and engineer, and Pauline Koch the daughter of a wealthy corn merchant. When Einstein was born, his 'extremely large and angular head' frightened his mother, who apparently was afraid that he was deformed, but after a short time the head took on its normal shape.

In 1880 the family moved to Munich, which was the capital of the Kingdom of Bavaria. They lived at first in a small rented house but after five years moved to a larger house, surrounded by big trees and a rambling garden. Then as now Munich was an overwhelmingly Catholic city. In Munich Einstein's father and his uncle Jakob founded Elektrotechnische Fabrik J. Einstein & Cie, a company that manufactured electrical equipment based on direct current. A year after the family's arrival in Munich Albert's sister Maria was born. She was called Maja by the family and, being only two years younger than Albert, she formed a close bond with him.

The Einsteins were non-observant Jews. Until 1871, only eight years before Einstein's birth, Jews didn't have the same rights as other Germans. Earlier in the century they had been forced to live in ghettos and had often been required to wear yellow badges (a practice revived by the Nazis). When Einstein was born his birth certificate recorded the fact that he was Jewish.

Einstein's mother loved music and she taught Albert both the piano and the violin, later arranging lessons. Starting at the age of five, Einstein spent three years at a Catholic elementary school in Munich. As religious instruction was compulsory he received that at home from a distant relative and he took it very seriously.

Starting at the age of eight, he then spent seven years at the Luitpold Gymnasium (now known as the Albert Einstein Gymnasium). At the Luitpold Gymnasium he started to receive instruction in algebra and geometry when he was thirteen. He had however already learned about those subjects from books during the previous summer. Einstein was not regarded as a good pupil, and when his father asked the headmaster what profession Albert should adopt the headmaster is supposed to have replied 'It doesn't matter, he'll never make a success of anything'.

In 1894 Hermann and Jakob's company lost a bid to supply the city of Munich with electrical lighting because they lacked the capital to convert their equipment from the direct current (DC) standard to the more efficient alternating current (AC) standard. The loss forced the sale of the Munich factory. In search of business, the Einstein family moved to Italy, first to Milan and a few months later to Pavia. When the family moved to Pavia, Einstein stayed in Munich to finish his studies at the Luitpold Gymnasium. Einstein clashed with the authorities and resented the school's regimen and teaching method. He later wrote that the spirit of learning and creative thought was lost in strict rote learning.

At the age of twelve Einstein was deeply religious, composing several songs in praise of God and chanting them on his way to school.

That changed when he read science books that appeared to contradict his religious beliefs.

Among the friends of Einstein's family was a medical student called Max Talmey. He was introduced by his elder brother, into 'the happy, comfortable and cheerful Einstein home, where I received the same generous consideration as he did'. Later, Talmey wrote an article called *Personal Recollections of Einstein's Boyhood and Youth* that was published in the journal *Scripta Mathematica*. In it we have a description of Einstein at the age of twelve:

> He was a pretty, dark-haired boy ... a good illustration ...against the theory of Houston Stewart Chamberlain and others, who try to prove that only the blonde race produces geniuses. He showed a particular inclination towards physics, and took pleasure in conversing on physical phenomena. I gave him therefore as reading matter, A. Bernstein's *Popular Books on Physical Science* and L. Buchner's *Force and Matter.*, two books that were then quite popular in Germany. The boy was profoundly impressed by them. Bernstein's work especially, which describes physical phenomena lucidly and engagingly, had a great influence on Albert, and enhanced considerably his interest in physical science.

Talmey also gave him *Critique of Pure Reason* by Immanuel Kant and *Cosmos: A Sketch of the Physical Description of the Universe* by Alexander von Humboldt. The latter was read in Germany at that time more than any other book except the Bible. The former espoused an absolute view of space and time which Einstein's later rejected through his Relativity theory.

At the end of December 1894 he travelled to Italy to join his family in Pavia, convincing the school to let him go by using a doctor's note. He liked Italy, writing later 'The people of northern Italy are the most civilised people I have ever met'. During his time in Italy he wrote a short essay with the title (translated to English) *On the Investigation of the State of the Aether in a Magnetic Field.*

As a German citizen, Einstein was obliged to report for conscription to the military before reaching the age of seventeen. Not wishing to do that, he renounced his citizenship in 1896. He was then stateless until he became a Swiss citizen in 1901.

In 1895, at the age of sixteen, Einstein took the entrance examinations for the Swiss Federal Polytechnic in Zurich (later the Eidgenössische Technische Hochschule, ETH). That was the best place in Europe to study science and engineering, outside of Germany. He hadn't done much preparation and he failed to reach the required standard in the language part of the exam, but he obtained exceptional grades in physics and mathematics. Because of those grades the Principal of the Polytechnic recognised Einstein's potential, and on the Principal's advice Einstein spent the following academic year at the cantonal school (gymnasium) in Aarau, Switzerland, completing his secondary schooling. This was a progressive school with an excellent reputation for science teaching, and Einstein spent one of the happiest and most fruitful years of his life there. He wrote the following (translated from the French, and so probably written in a French class)

My plans for the future

A happy man is too content with the present to think much about the future. Young people, on the other hand, like to occupy themselves with bold plans. Furthermore, it is natural for a serious young man to gain as precise an idea as possible about his desired aims.

If I were to have the good fortune to pass my examinations, I would go to the [ETH] in Zurich. I would stay there for four years in order to study mathematics and physics. I imagine myself becoming a teacher in those branches of natural science, choosing the theoretical part of them.

Here are the reason which led me to this plan. Above all, it is [my] disposition for abstract and mathematical thought, [my] lack of imagination and practical ability. My desires have also inspired in me the same resolve. That is quite natural: one always likes to do the things for which one has the ability. Then there is also a certain independence in the scientific profession, which I like a great deal.

Later in life Einstein wrote of his time at the ETH (contradicting somewhat his French essay) the following

I had excellent teachers ... so that I really could have gotten a sound mathematical education. However, I worked most of the time in the physical laboratory, fascinated by the direct contact with experience. The balance of the time, I used in the main to study at home.

While lodging with the family of professor Jost Winteler, Einstein fell in love with Winteler's daughter Marie. (Albert's sister Maja later married Winteler's son Paul.)

In September 1896 he passed the Swiss school-leaving exam, with mostly good grades including the top grade in physics and mathematical subjects. At seventeen he enrolled in the four-year mathematics and physics teaching diploma program at the Zurich Polytechnic. Marie Winteler, who was a year older, moved to Olsberg, Switzerland for a teaching post. At the polytechnic he became friends with his fellow-student Marcel Grossmann, whose family had emigrated from Budapest. Grossmann gave Einstein his lecture notes to copy because Einstein often missed classes and had incomplete notes.

Einstein was absent-minded, frequently having to knock up his landlady because he'd forgotten his key. The absent-mindedness probably came from his constantly thinking physics. One of his students recalled that he once stood under a lamp post during a snowstorm, and handed his umbrella to a student while he wrote down formulae for ten minutes.

Einstein's future wife, Mileva Maric, also enrolled at the Polytechnic that year. She was born in 1875, and so was a bit older than Einstein as were most of the class. She was the daughter of a Serbian peasant from Titel in southern Hungary and was handicapped by a limp. Still, she had persevered successfully with her education. Mileva Maric was the only woman among the six students in the mathematics and physics section of the teaching diploma course. Over the next few years Einstein and Mileva's friendship developed into romance and they read books together on extra-curricular physics in which Einstein was taking an increasing interest. In 1900 Einstein was awarded the Zurich Polytechnic teaching diploma, but Mileva failed the examination with a poor grade in one of the mathematics components.

After graduating in 1900 Einstein wanted to continue to a doctorate but was not accepted. His teachers felt that his attitude was disrespectful, and they remembered that his attendance in class had been sporadic. They could not have been aware that he was doing, instead, a lot of private study. Einstein then did some irregular teaching but it was two years before he became fully employed.

Einstein acquired Swiss citizenship in February 1901, which he retained for the rest of his life. A period of military service was usually compulsory for Swiss men but he was not conscripted because he had flat feet and varicose veins. With the help of Marcel Grossman's father he secured in 1902 a job in Bern at the Federal Office for Intellectual Property, which he retained for seven years. Einstein evaluated patent applications for a variety of devices, including a gravel sorter and an electromechanical typewriter. His friend Max Talmey visited him and afterwards wrote

> He lived in a small poorly furnished room. I learned that he had a hard life struggle with the scant salary of an official at the Patent Office.

Much of his work at the patent office related to questions about transmission of electric signals, which may have influenced his later research.

In 1902, Einstein started a discussion group with some friends that he had met in Bern. They self-mockingly named it 'The Olympia Academy' and it met regularly to discuss science and philosophy. One of the authors whose works they read was the eighteenth century Scottish philosopher, historian and economist David Hume, who wrote 'A wise man proportions his belief to the evidence'. Another was the nineteenth century Austrian physicist and philosopher Ernst Mach, after whom the Mach number measure of speed is named. A third was the nineteenth century French mathematician, physicist, engineer and philosopher Henri Poincaré, whose work foreshadowed Einstein's Special Relativity.

The discovery and publication in 1987 of an early correspondence between Einstein and Mileva Maric revealed that they had a daughter, called Lieserl in their letters. She was born in early 1902 in Novi Sad where Mileva was staying with her parents. Mileva returned to Switzerland without the child, whose real name and fate are unknown. Einstein probably never saw his daughter. The contents of a letter to Mileva in September 1903 suggest that the girl was either given up for adoption or died of scarlet fever in infancy.

Einstein and Mileva married in January 1903. The marriage was welcomed by her parents, but not by Einstein's mother. It seems that this was because Mileva was a Christian. In May 1904 their first son Hans Albert was born in Bern. He graduated at the ETH in Zurich, to become an engineer living in Dortmund. In 1938 he fled Nazi Germany, to become a professor of hydraulic engineering in UC Berkeley. He died in 1973 at the age of sixty-nine. Their second son, Eduard was born in Zurich in July 1910. Eduard, whom his father called 'Tete' (for petit), had a breakdown at about age twenty and was diagnosed with schizophrenia. His mother cared for him but he was also committed to asylums for several periods, finally being committed permanently after her death in 1948. He died in 1965 at the age of fifty-five.

On 30 April 1905, at the age of twenty-six, Einstein submitted a doctoral thesis to the University of Zurich, with Alfred Kleiner, Professor of Experimental Physics, as his doctoral advisor. As a result Einstein was awarded in 1905 a doctorate by the University of Zurich. (Einstein's initial advisor had been Heinrich Friedrich Weber, another experimental physicist. They fell out, and Einstein complained that his lectures were fifty years out of date and ignored Maxwell's equations.)

Later in 1905 Einstein published four ground-breaking papers. There have been claims that Mileva collaborated with Einstein on these papers, but historians of physics who have studied the issue find no evidence that she made any substantive contributions. We do know though that they celebrated the publication of the papers together, getting, uncharacteristically, quite drunk. We know it because Einstein's friend Conrad Habicht received a postcard saying 'Both of us, alas, dead drunk under the table'.

There's an anecdote about the 1905 papers, which illuminates the relation between Einstein, his father and his mother. Two of the 1905 papers covered Special Relativity, and when they came out Einstein wrote letters to his father explaining that theory. Seeing them, Einstein's mother asked him to include her in the discussion, saying 'please make it simple enough for a corn merchant's daughter'. Einstein's sister Liserl reports, however, a scene suggesting that his mother understood quite a lot about his theory. At Christmas in 1907, she came across her mother explaining Relativity to an aunt in lucid terms, and she writes that her mother cut short her explanation when Einstein and his father walked in, saying: 'He (pointing at my father) and he (pointing at my brother) are frolicking in a Platonic realm of pure ideas, while I am standing in a Württemberg corn field. I'm up to my ears in corn!'

Einstein's ground-breaking papers were not at first widely appreciated. One person who did take them seriously though was the German physicist Max Planck. Planck wrote to Einstein in 1907 'As long as the proponents of the Principle of Relativity constitute such a modest little band as is now the case, it is doubly important that they agree among themselves'.

The work so far mentioned was done in Einstein's spare time, while he was putting in a forty-eight-hour week at the Patent Office. The same is true of some other work that he did at the same time, marking the beginning of his journey toward the theory of General Relativity.

In 1903 he had tried to become a lecturer at the University of Bern and he tried again in 1907, with now seventeen published papers. He was told that his application was denied until he produced his Habilitation thesis. (A Habilitation thesis is a more advanced work than a doctoral thesis, and is required in many European countries before one may give lectures.)

In September 1907 Johannes Stark invited Einstein to contribute a review of Relativity, to a journal that he had founded. (Stark was an eminent experimental physicist, discovering what is called the Stark Effect and receiving the Nobel Prize in 1919.) Einstein agreed but he wrote that '[I am] not in a position to acquaint myself with *everything* published on this topic, because the library is closed during my free time'. Even so he submitted the review in late November. The review included Special Relativity, and as much of General Relativity as Einstein had by that time figured out. Stark welcomed the article at the time but in the 1930s he founded the anti-Semitic movement Deutsche Physik (German Physics), which condemned Einstein, and even the entire enterprise of theoretical physics, as Jewish.

In January 1908 Einstein, still at the Patent Office, applied for a teaching post at a high school in Zurich. One of twenty-one applicants he was not even short-listed! Later in that year though, he was appointed lecturer at the University of Bern. The following year, after he gave a lecture on electrodynamics and the Relativity Principle at the University of Zurich, Alfred Kleiner recommended him to the faculty for a newly created professorship in theoretical physics. Einstein was appointed associate professor in 1909.

Einstein was very popular as a lecturer, being very clear even though he rarely used notes. A friend wrote

> He went to a great deal of trouble over his first lecture in order to help students, He kept stopping and asking whether he was understood. In the pauses [between lectures] he was surrounded by students who wanted to ask questions, which he answered in a most patient and friendly way.

After the weekly physics colloquium he would take students to the cafe, and bring them home to discuss physics.

In April 1911 Einstein became a professor at the German Charles-Ferdinand University in Prague. To do so he accepted Austrian citizenship in the Austro-Hungarian Empire. During his Prague stay he wrote eleven scientific papers. Just as he had done in Zurich, he organised a weekly colloquium, with discussions afterward. The Einsteins left after one year though, to return to Zurich. This was largely because Einstein found his duties in Prague too onerous, including some routine laboratory work. Mileva welcomed the return because she had found it difficult to manage the change.

From 1912 until 1914 he was professor of theoretical physics at the ETH in Zurich. During those two years the Einsteins had several distinguished visitors. One was the Polish physicist and chemist Marie Curie, who pioneered the study of radioactivity, with her two daughters. They had shown the Einsteins the ropes when they visited Paris in March 1913.

In 1912, while on a trip to Berlin, Einstein began an affair with Elsa Lowenthal. Elsa was his first cousin on his mother's side, and his second cousin on his father's side.

In July 1913 he was visited by Max Planck and the German chemist Walther Nernst. They asked him to join the Prussian Academy of Sciences in Berlin, and also offered him the post of director at the Kaiser Wilhelm Institute for Physics which was soon to be established. (Membership in the academy included paid salary and professorship without teaching duties at the University of Berlin.) He was officially elected to the academy on 24 July and he agreed to move to Berlin the next year.

His decision to move to Berlin was influenced by the prospect of living near Elsa Lowenthal. In Berlin Einstein wanted Mileva and himself to be separated. She at first refused, upon which he drew up a list of outrageous conditions that she must fulfil if they were to stay together. One of them read 'You are neither to expect intimacy from me, nor to reproach me in any way'. Mileva accepted them but they split up anyway. Mileva kept the children and receiving an annual sum which, when it began, represented nearly half of Einstein's income. Later he also gave Mileva his Nobel Prize money. Einstein saw little of his children after the separation.

He joined the academy, and thus the University of Berlin, on 1 April 1914. As World War I broke out that year the plan for Kaiser Wilhelm Institute for Physics was aborted. The Institute was finally established on 1 October 1917 with Einstein as its Director. In 1916, Einstein was elected president of the German Physical Society and served for two years. He was a citizen of the Free State of Prussia, within the Weimar Republic, from 1918 to 1933.

When World War I broke out, ninety-three of Germany's leading scientists signed a pro-war 'Manifesto of the 93'. Einstein joined with George Nicolai, the Professor of Physiology in the University of Berlin, in producing an anti-war manifesto but they could get only two others to sign it. One of them had also signed pro-war manifesto – clearly not a person who read the small print! Their manifesto remained unpublished for many years.

In 1915 Einstein published his theory of General Relativity. A prediction of General Relativity was tested soon after the theory came out. It concerned the deflection of light from a star when it passes close to the Sun. Based on calculations that he had already made in 1911, Einstein found that the deflection should be twice as big as the one given by Newton's theory of gravity.

To test Einstein's prediction, one needs a total eclipse of the Sun when there's a star very near the Sun's edge. Such an eclipse occurred in 1919, and Einstein's prediction was tested in an expedition to the island of Principe off the west coast of Africa, led by the Astronomer Royal Frank Watson Dyson, and the English astronomer, physicist and mathematician Arthur Eddington. (It was also tested in an expedition to Brazil by Andrew Claude de la Cherois Crommelin and Charles Davidson from the Royal Observatory at Greenwich.)

Having made the test, Dyson and Eddington announced that the prediction had been confirmed. There has since been some doubt as to whether the observations were really accurate enough to justify the announcement, but be that as it may, the story was published in the international media, making Einstein world famous. For example, On 7 November 1919, the leading British newspaper *The Times* printed a banner headline that read: 'Revolution in Science — New Theory of the Universe –- Newtonian Ideas Overthrown'.

Eddington later pointed out that the sighting of a star near the edge of the sun during an eclipse is a rare occurrence, and said that if General Relativity had been proposed at another time we might have had to wait thousands of years for this prediction to be verified. As we shall see, the second statement is no longer true, now that we have space travel.

Eddington had been one of the few people who understood Einstein's General Relativity. Also, being a pacifist, he was prepared to pursue it during World War I despite its German origin. There's an amusing tale concerning Eddington and Relativity. At some point, Eddington was approached by the Polish-American physicist Ludwick Silberstein, who considered himself an expert on General Relativity.

Silberstein suggested to Eddington that he was one of only three people who understood Einstein's General Relativity (the others in Silberstein's mind being himself and Einstein). When Eddington didn't immediately reply, Silberstein said 'don't be so shy', but Eddington replied 'no, I was wondering who the third person might be'!

On 14 February 1919, Einstein and Mileva were divorced having lived apart for five years. In the same year Einstein married Elsa Lowenthal. They lived with Elsa's two daughters from a previous marriage. Elsa wrote that she fell in love with Einstein 'because he played Mozart so beautifully on the violin'. For his part, Einstein's regard for Elsa was very different from the way that he had regarded Mileva. She was more of a mother figure, speaking the same dialect as him and liking the same food. Einstein had affairs with other women, but Elsa was always there, protecting this now-famous man from unwanted attention and taking care of him generally. In letters revealed in 2015, Einstein wrote to his early love, Marie Winteler, about his marriage and his still-strong feelings for Marie. In 1910, while his wife was pregnant, he wrote to her that 'I think of you in heartfelt love every spare minute and am so unhappy as only a man can be'. Einstein spoke about a 'misguided love' and a 'missed life' regarding his love for Marie.

Life in Germany after World War I was difficult. This was largely because of severe war reparations imposed by the Allies. Einstein wrote the following to his mother, who was bedridden with cancer in Switzerland.

> We have to relinquish a room [regulations required that it be rented out]. Starting tomorrow, the elevator won't work ... much shivering lies ahead of us in winter.

In 1919 Einstein's mother joined them. Also in that year, Einstein agreed to support Zionists, who wanted to establish a Jewish state in Palestine. (Later, in 1930, he said that he would support them only if they tried sincerely to make peace with the Arabs.) Einstein's support for Zionism made him unpopular in Germany. His view that Germany had committed war crimes during World War I also made him unpopular.

In 1921 Einstein visited the USA. His English was very poor, but Elsa's was good. When he arrived in New York reporters noticed that he was holding his violin, which he often played to take a break from thinking. They asked about Relativity without receiving much enlightenment, and then they asked for Einstein's thoughts on American culture. He was amazed how well even working-class women dressed. He condemned Prohibition and could not imagine banning tobacco. He liked films, though he thought that they were at an early stage of development, he thought bathrooms were marvellous. After half an hour he ended the interview, though Elsa continued to answer questions in fluent English.

Einstein was visiting the USA chiefly to raise funds for the planned Hebrew University of Jerusalem, but he did give lectures in German which were well attended because of his fame. After one of them he's supposed to have naively said, that he hadn't realised how many Americans were interested in tensor analysis! Einstein published in July 1921 *My First Impression of the U.S.A.* In it he wrote: 'What strikes a visitor is the joyous, positive attitude to life . . . The American is friendly, self-confident, optimistic, and without envy'.

In 1922 Einstein was awarded the Nobel Prize for Physics. In the same year he and Elsa visited Paris, and then Japan. On arrival in Japan he gave a lecture, where according to the *Japan Weekly Chronicle* 'The hall was filled with scholars, teachers and students. Some women were present too'. This lecture, and subsequent ones, were of course translated from German into Japanese, which meant that they took a long time. The Einsteins were given the great honour of being introduced to the Emperor, and to the Empress with whom Einstein conversed in French.

In a letter to his sons he described his impression of the Japanese as being modest, intelligent, considerate, and having a true feel for art. Einstein also visited China, which he viewed less favourably; in his diary he wrote that the Chinese were

> an industrious, filthy people, who don't sit on benches while eating but squat like Europeans do when they relieve themselves out in the leafy woods. All this occurs quietly and demurely. Even the children are spiritless and look obtuse. It would be a pity if these Chinese supplant all other races. For the likes of us the mere thought is unspeakably dreary.

When contrasting these comments, one should remember that Einstein visited a Japan that had been peaceful for several generations, but a China that was in turmoil. Also, that sitting on something other than the ground was not the practice of people in most countries at the time, and that one could sit on the ground only where it was suitable for that purpose.

On his return voyage he visited Palestine, where he was greeted with a cannon salute upon arriving at the home of the British high commissioner, Sir Herbert Samuel. During one reception the building was stormed by people who wanted to see him. In Einstein's talk to the audience, he said that he was happy that the Jewish people were beginning to be recognized as a force in the world.

Returning to Berlin he was troubled by the rise of Hitler's National Socialists, but stayed on. He worked well there, and enjoyed life there generally. He sailed in the lakes that surrounded Berlin, and also got know the violinist Fritz Kreisler and the pianist Arthur Schnabel, two of the leading musicians of their day.

Einstein was a leader in establishing the Hebrew University of Jerusalem, which opened in 1925, and he served on the Board of Governors from 1925 to 1928. From 1922 to 1932, Einstein was also involved with the International Committee on Intellectual Cooperation of the League of Nations, in Geneva.

There's an amusing story about Einstein and the Queen of Belgium. On a visit to Leiden in 1929 he received an invitation to visit the Queen (King Albert was away in Switzerland). Queen Elizabeth, formerly Princess Elizabeth of Bavaria, played the violin and set up musical afternoons with Einstein. He often visited subsequently. Once, looking as usual somewhat dishevelled, he asked a small cafe for the use of their phone, and alarmed them by asking to be put through to the Queen!

In 1930 Einstein visited England, where he spoke with Arthur Eddington. He had said to Eddington in 1926 that it would be worth learning English just to talk with him! At the end of that year he made a visit to the USA, which was supposed to be for two months as a research fellow at the California Institute of Technology. In New York, he attended a performance of *Carmen* at the Metropolitan Opera, where he was cheered by the audience upon arriving. In California he met Caltech president and Nobel laureate, Robert A. Millikan, but that didn't go very well, because Millikan was patriotic and militaristic whereas Einstein was an internationalist and a pacifist. During an address to Caltech's students, Einstein said that science often did more harm than good. In California, Einstein also met Edwin Hubble, whose observations had recently demonstrated that the universe is expanding. Einstein had accepted this, giving up his static model of the universe.

Returning to New York, Einstein's aversion to war led him to befriend author Upton Sinclair and film star Charlie Chaplin, both noted for their pacifism. He had an instant rapport with Chaplin, who invited

Einstein and his wife to his home for dinner. Chaplin said that Einstein's outward persona, calm and gentle, seemed to conceal a 'highly emotional temperament', from which came his 'extraordinary intellectual energy'. Chaplin also invited Einstein and Elsa to join him for the premier of his film *City Lights*, as his guests.

In the early 1930s Einstein made one of several visits to his sister Maria. She was living in Colonnata near Florence, with her lawyer husband Paul Winteler. When at Colonnata, Einstein also visited his cousin Robert and family at Troghi, near Florence. A watercolour by Robert's young niece, Lorenza Mazzetti, has the following caption (translated to English).

> This is a portrait of Uncle Robert's cousin, Maja's brother. His name is Albert Einstein. He lives in America and when he is there, he works as a scientist, and when he comes here, he goes on the swing.

Robert's wife and two adult daughters were later to be tortured and shot by the occupying German SS. Robert escaped that fate, but he was so scarred by the experience that he committed suicide.

In February 1933, while Einstein was on another visit to the USA, Hitler became Germany's Chancellor, and Einstein decided not to return to Germany. A German magazine later included him in a list of enemies of the German regime with the phrase, 'not yet hanged', offering $5,000 for his head.

In late July 1933 he went to England for about six weeks at the invitation of British naval officer Oliver Locker-Lampson. To protect Einstein, Locker-Lampson supposedly had people watching over him at his secluded cottage outside London, and a photo of them carrying shotguns and guarding Einstein was published in the *Daily Herald* on 24 July 1933. Actually, the 'guards' were two female secretaries and a farm hand, who had been given rifles for the benefit of the photographer!

Locker-Lampson took Einstein to meet Winston Churchill at his home, and later to meet Austen Chamberlain and former Prime Minister Lloyd George. Einstein asked them to help bring Jewish scientists out of Germany. Churchill responded immediately. He sent the British physicist Frederick Lindemann to Germany, who sought out Jewish scientists for placement in British universities. Einstein later contacted leaders of other nations, including Turkey, requesting placement of unemployed German-Jewish scientists. More than a thousand Jewish refugees were eventually invited to Turkey.

In 1933, Einstein returned to the USA, and took up a position at the Princeton Institute for Advanced Study. The Princeton Institute had become something of a refuge for Jewish scientists fleeing Nazi Germany, because the Universities suffered from a Jewish quota for both staff and students, which lasted until the late 1940s. Einstein's affiliation with the Institute for Advanced Study would last until his death in 1955. His research was not proving successful though, and in 1934 he wrote to his friend, the Queen of Belgium: 'I have locked myself into quite hopeless scientific problems – the more so since, as an elderly man, I have remained estranged from the society here...'

In 1935 Einstein applied for, and was granted, permanent residency in the USA. In the same year Einstein's wife Elsa was diagnosed with heart and kidney problems, and she died in December 1936. In 1938 Einstein was joined by Elsa's daughter Margot, and also by Einstein's younger sister Maria who was fleeing Mussolini's Italy. Knowing that he would not be allowed entry to the USA on health grounds, Maria's husband took refuge with relatives in Geneva. They were never to see each other again.

In July 1939, a few months before the beginning of World War II in Europe, Einstein was visited by the Hungarian-born physicist Leo Szilard and the Hungarian-born theoretical physicist, mathematician and engineer Eugene Wigner. They explained the possibility of atomic bombs, and asked him to write a letter, with Szilard, to President Franklin D. Roosevelt, warning him that Germany might develop atomic bombs, and suggesting that the USA should do the same. Einstein wrote the letter and met Roosevelt, and he also used his connection with the Belgian Queen Mother to get access with a personal envoy to the White House's Oval Office. Partly as a result of these actions, the USA became the only country to successfully develop a nuclear bomb during World War II. A year before his death, Einstein was to say to the American chemist, biochemist and peace activist Linus Pauling, 'I made one great mistake in my life–when I signed the letter to President Roosevelt recommending that atom bombs be made; but there was some justification–the danger that the Germans would make them ...'.

In 1940 Einstein became a US citizen, while retaining his Swiss citizenship. In 1944 he made a contribution to the US war effort by hand writing his 1905 paper on Special Relativity and putting it up for auction. It raised $6,000,000, the manuscript today being in the Library of Congress.

In 1946 Einstein's sister Maria suffered a stroke and became bedridden. After that Einstein spent part of every evening reading to her, until her death in 1951. Among the books was James Frazer's *Golden Bough*, which describes the development of thought from magic to science.

On 17 April 1955, Einstein experienced internal bleeding caused by the rupture of an aortic aneurysm. He refused surgery, saying: 'I have done my share, it is time to go', and he died in Princeton Hospital early the next morning at the age of seventy-six. He left his scientific papers to the Hebrew University of Jerusalem. One week before his death, he had signed a letter to Bertrand Russell, agreeing that his name should go on a manifesto urging all nations to give up nuclear weapons.

Throughout his life Einstein published hundreds of books and articles. He published more than 300 scientific papers and 150 non-scientific ones. On 5 December 2014, universities and archives announced the release of Einstein's papers, comprising more than 30,000 unique documents. Nowadays the word 'Einstein' is synonymous with 'genius'.

Einstein's large output contains many oft-quoted passages. Here is one, from *On the Methods of Theoretical Physics* published in 1933

> Our experience hitherto justifies us in believing that nature is the realization of the simplest conceivable mathematical ideas. I am convinced that we can discover by purely mathematical constructions the concepts of the laws connecting [mathematics and physical reality] with each other, and furnish the key to the understanding of natural phenomena.... In a certain sense, therefore, I hold it true that pure thought can grasp reality, as the ancients dreamed.

and here is another, translated from an address to the Prussian Academy of Sciences in 1921 titled *Geometry and Experience*

> As far as the laws of mathematics refer to reality, they are not certain, and as far as they are certain they do not refer to reality.

To round out this account of Einstein, it may be helpful to mention some of his habits and opinions. Except for smoking, his habits were moderate. He drank little alcohol, but still enjoyed a glass of wine or brandy. He once looked in astonishment at a colleague who turned down a glass of wine, saying 'One should not neglect the pleasures that nature provides'.

He offered, or was called on to give, opinions on matters often unrelated to theoretical physics or mathematics. Some of these concerned music. One biographer wrote the following

> Few aspects of Einstein's life and personality are cited with such regularity, in nearly identical phrases and commentary, as his devotion to and love of music. Every biographical account tells the same story. His mother, a pianist of reasonable amateur proficiency, wanted her son to play the violin. It is speculated that she wished to have a partner in the family for *Hausmusic*. Musical culture in the home, which meant chamber music, usually with piano, was a highly prized symbol of successful middle-class acculturation, in late 19th-century German-speaking Europe. It signaled *Bildung*, a sign of status and achievement particularly prized by assimilated urban Jews.

Einstein later said that he taught himself to play without 'ever practicing systematically', deciding that 'love is a better teacher than a sense of duty'. At age seventeen, he was heard by a school examiner in Aarau, playing a Beethoven violin sonata. The examiner stated afterward that his playing was remarkable and revealed 'great insight'. What struck the examiner, writes Borstein, was that Einstein 'displayed a deep love of the music, a quality that was and remains in short supply. Music possessed an unusual meaning for this student'.

Einstein wrote that Mozart's music 'was so pure that it seemed to have been ever-present in the universe, waiting to be discovered by the master'. He also revered Bach, but disliked Wagner, writing 'I see his lack of architectural structure as decadence. Moreover, to me his musical personality is indescribably offensive so that for the most part I can listen to him only with disgust.' About Debussy, he wrote in a 1939 questionnaire: 'I feel that Debussy is delicately colorful but shows a poverty of structure. I cannot work up great enthusiasm for something of that sort'.

He played regularly until he was prevented by old age, rarely going anywhere without his violin, which he always called Lina irrespective of which instrument he had at the time. He wrote in old age 'If I were not a physicist, I would probably be a musician. I often think in music. I live my daydreams in music. I see my life in terms of music... I get most joy in life out of music.' He also wrote more specifically 'most joy in my life has come to me from my violin'.

In 1931, while engaged in research at the California Institute of Technology, he visited the Zoellner family conservatory in Los Angeles, where he played some of Beethoven and Mozart's works with members of the Zoellner Quartet. Near the end of his life, when the young Juilliard Quartet visited him in Princeton, he played his violin with them, and the quartet was 'impressed by Einstein's level of coordination and intonation'. Things did not always go well though. When Einstein was playing in a quartet with his friend Fritz Kreisler, and missed yet another entrance, Kreisler exclaimed 'What's the matter professor? Can't you count?'

He joined National Association for the Advancement of Colored People (NAACP) in Princeton, and said that racism was America's 'worst disease', being 'handed down from one generation to the next'. He corresponded with civil rights activist W. E. B. Du Bois and was prepared to testify on his behalf during his trial in 1951. As a result, the judge decided to drop the case. In 1946 Einstein visited Lincoln University in Pennsylvania, a historically black college, at which he was awarded an honorary degree. (Lincoln was the first university in the United States to grant college degrees to African Americans.) Einstein gave a speech about racism in America, saying 'I do not intend to be quiet about it.' A resident of Princeton recalls that Einstein had once paid the college tuition for a black student.

Einstein was in favor of socialism and critical of capitalism. He explained this in essays such as *Why Socialism?*. He strongly advocated the idea of a democratic global government that would check the power of nation-states. By the time Einstein died, the FBI had a secret dossier on Einstein which was 1,427 pages long!

Regarding religion, Einstein stated that he believed in the pantheistic God of the seventeenth century philosopher Baruch Spinoza. He did not believe in a personal God who concerns himself with fates and actions of human beings, a view which he described as naive. He clarified however that, 'I am not an atheist', preferring to call himself an agnostic, or a 'deeply religious nonbeliever'. When asked if he believed in an afterlife, Einstein replied, 'No. And one life is enough for me'.

10.2 SCIENCE

In 1894, when Einstein was fifteen, Albert Michelson said the following at the opening of the Ryerson Physical Laboratory at the University of Chicago

> The most important fundamental laws and facts of physical science have all been discovered, and these are now so firmly es-

tablished that the possibility of their ever being supplemented in consequence of new discoveries is exceedingly remote.

The reaction of those present to this negative view of their endeavors is not recorded!

Before getting to Einstein's Relativity I'll mention some of his other work. It began in 1900 when he was twenty-one. Over the next five years, he published four papers on thermodynamics and statistical mechanics. In 1905 he was awarded a doctorate by the University of Zurich. His seventeen-page thesis was on what's called Brownian motion. This is the random motion of tiny particles that are suspended in a liquid or gas. It's named after Robert Brown, the Scottish botanist and palaeobotanist who observed it in the early nineteenth century, but it has a long history. It was first described by the Roman poet and philosopher Lucretius in the first century BC, who wrote the following

> Observe what happens when sunbeams are admitted into a building and shed light on its shadowy places. You will see a multitude of tiny particles mingling in a multitude of ways... their dancing is an actual indication of underlying movements of matter that are hidden from our sight... It originates with the atoms which move of themselves [i.e., spontaneously]. Then those small compound bodies that are least removed from the impetus of the atoms are set in motion by the impact of their invisible blows and in turn cannon against slightly larger bodies. So the movement mounts up from the atoms and gradually emerges to the level of our senses, so that those bodies are in motion that we see in sunbeams, moved by blows that remain invisible.

As Lucretius stated, Brownian motion is indeed caused by the collision of the atoms or molecules of the liquid. The existence of atoms was not universally accepted even by Einstein's day, even though evidence for them had been presented in the early nineteenth century by the English chemist, physicist and meteorologist John Dalton. Einstein did calculations, which gave a detailed description of Brownian motion. He found that the motion comes, not from the collision of a single atom, but of that of many atoms which happen to be moving in more or less the same direction when they strike one of the particles. His calculation allowed him to determine for the first time, the size of the atoms of which the liquid is composed. Einstein's calculation was verified in an experiment performed in 1908 by the French physicist Jean Baptiste Perrin, which proved beyond doubt the existence of atoms.

In 1905, Einstein published the four ground-breaking papers mentioned earlier. One was on what's called the photoelectric effect, in which light shone onto a metal causes electrons to be emitted. To explain what was observed, Einstein proposed that light (or any electromagnetic wave) is composed of elementary particles. We now call those particles photons. He proposed that the energy of each photon would be equal to the frequency of the wave, times a number called Planck's constant. (By frequency, one means the number of oscillations which take place in one second.) Max Planck had proposed this relation between energy and frequency about five years earlier, which is why the number's called Planck's constant. Planck, though, had not proposed the existence of actual particles.

In 1907, and again in 1911, Einstein developed a theory of specific heats, which used the idea that energy comes in discrete bundles. (The specific heat of a substance can be defined as the amount of energy required, to raise one kilogramme of the substance by one degree Celsius.) His theory resolved a paradox of nineteenth century physics, that specific heats were often smaller than could be explained by any existing theory.

The idea that energy comes in bundles, was fully developed only in the 1920s. The bundles were then called quanta, and the resulting theory was called quantum physics. We have just seen that quantum physics allows one to think of a beam of photons as a wave, but quantum physics goes much further. It says that we can think of a beam of electrons, or any other tiny particles, also as a wave. Crazy! Things like that led Richard Feynman, one of the pioneers of quantum physics, to say that nobody really understands quantum physics.

Einstein's second paper of 1905 was on Brownian motion. The third paper introduced Special Relativity. The fourth paper introduced the famous relation $E = mc^2$ (energy = mass times the speed of light squared), which relates the energy of a stationary object to its mass.

In 1910 Einstein developed the theory of critical opalescence, an effect which causes a fluid to look milky under the right conditions. He related critical opalescence to Rayleigh scattering, by virtue of which the sky is blue. In 1918 Einstein developed a theory of the process by which atoms emit and absorb electromagnetic radiation. That is the theory behind lasers.

When Einstein received the 1921 Nobel Prize in Physics, it was 'for his services to Theoretical Physics, and especially for his discovery of the law of the photoelectric effect'. Relativity was still considered somewhat controversial, and was therefore not cited. Also the citation concerning the photoelectric effect does not treat the cited work as an explanation, but merely as a discovery of the law, because the idea of photons was considered outlandish.

Undeterred by such scepticism, Einstein and the Bengali physicist Satyendra Nath Bose pursued the idea of photons. The ideas that they came up with are now widely used. In particular, they are used in the construction of the MRI and NMR scanners used in medicine.

In 1935 Einstein with the Russian-American physicist Boris Podolsky and the American (later Israeli) physicist Nathan Rosen put forward what is now known as the EPR paradox. This is a surprising consequence of quantum physics, which they regarded as evidence that quantum physics is incorrect. Nowadays, the consequence is simply accepted.

Even though he had been one of the founders of the subject, Einstein became unhappy about quantum physics as it developed. That's because, as I mentioned at the end of the chapter on Newton, quantum physics doesn't allow definite predictions about what will happen. In a letter to the German physicist and mathematician Max Born written in 1925 he wrote that he was 'convinced that God does not throw dice'. In 1927, Einstein attended a conference in Brussels, along with 28 other physicists, more than half of them Nobel prizewinners. By this time, quantum physics was beginning to be accepted, and when Einstein repeated his view that 'God does not play dice with the universe', the Danish physicist Neils Bohr replied 'don't tell God what to do'!

Finally, we come to Einstein's theories of Special Relativity and General Relativity. Einstein saw that Maxwell's equations presented a problem for his Principle of Relativity. The problem is evident, just from the fact that Maxwell's equations give the speed of light. The combination of Maxwell's equations and the Principle of Relativity, demands that the speed of light is independent of the motion of the observer who is measuring that speed. This seems to make no sense, because one would think that the speed measured by an observer moving in the direction of the light, would become smaller as the observer moved faster.

To make the Principle of Relativity compatible with Maxwell's equations, Einstein in 1905 put forward his theory of Special Relativity. This theory takes seriously the fact that a distance should be defined by some measuring rod, while a time interval should be defined by some clock. According to Special Relativity, the times and distances, measured by clocks and rods, depend on the motion of those clocks and rods. The dependence is such, that the speed of light that they measure is always the same, regardless of their motion.

To see how clocks behave according to Special Relativity, imagine a pair of identical twins. One stays on the earth while the other makes a very fast trip outwards into space for six months, and then turns round, to return to earth at the same speed. According to Special Relativity, the earthbound twin will be older than the one who made the journey. If the speed of the journey is close to the speed of light, this effect could be huge, and the travelling twin might land back to earth in the far future with his twin long dead, like Rip Van Winkle awakening from his long sleep!

From Special Relativity, Einstein deduced the relation $E = mc^2$. This relation means that if an object radiates energy, its mass decreases. Since the radiation of energy from objects is a familiar thing, one might wonder why the relation had not been found long ago. The answer is, that an object has to radiate a huge amount of energy before its mass decreases by a noticeable amount. To clarify that, Einstein wrote that a 100 watt light bulb switched on for 100 years, has its mass decreased by only 1/300th of a gram. Conversely though, if one can convert a *significant* amount of mass into energy, the yield will be enormous. That can indeed be achieved in nuclear reactions, which allows the construction of nuclear power stations and unfortunately also of nuclear weapons. The conversion of mass into energy also makes stars shine.

The most outstanding success of Special Relativity, is that its description of space and time is adopted in what is called the Standard Model of particle physics. The Standard Model agrees with observation, and encapsulates almost everything we know about particles and their interactions. Einstein's misgivings notwithstanding, quantum physics is a fundamental ingredient of the entire Standard Model.

Now I come to General Relativity. One of the main successes of General Relativity, and the thing that set Einstein on the path towards that theory, concerns the orbit of the planet Mercury. Mercury is the planet closest to the Sun, and the one whose orbit is the least circular. The point on the orbit which is closest to the Sun is called the perihelion of Mercury, and according to observation it moves slowly with time. Newton's theory of gravity does predict some movement of the perihelion, due to the effect of the other planets, but it's less than the movement which is observed. Einstein knew that, and it was one of his main motivations to search for an alternative theory of gravity. He found General Relativity which, happily, gives an amount of precession that agrees exactly with observation.

To see how Einstein arrived at General Relativity, recall first that the Relativity Principle, invoked by Special Relativity, refers to measurements by rods that are rigid and by clocks that register the true time. To go further, Einstein allowed the rods to be deformed, and the clocks to run erratically. He demanded that even then, the laws of physics should be such that they remained valid. Crazy as that idea sounds, Einstein made it work, arriving thereby at the theory of General Relativity. The most important result of General Relativity is encapsulated in a single equation, called the Einstein field equation.

Einstein developed General Relativity in stages, starting in 1907 and ending finally in 1915. To do so, he had to learn a new branch of mathematics called non-Riemannian geometry. He wrote that this gave him a new respect for mathematics, which he had previously regarded as rather trivial. (Einstein was introduced to non-Riemannian geometry by his old friend Marcel Grossmann.) General Relativity contains within it, a theory of gravity which goes beyond Newton's theory, and which agrees with a host of astronomical observations:

- The orbits of the planets, in particular the orbit of Mercury.

- The deflection of electromagnetic radiation by the sun.

- The gravitational time delay, whereby the arrival of electromagnetic radiation passing close to the sun is delayed. This effect has been measured using a radio signal from a space station on the other side of the sun, and so has the previous effect.

- The focussing of light from a distant object, by an object along the line of sight.

- The fact that the gravitational acceleration of an object at a given location, is independent of its mass and composition (equivalence principle).

- The spiraling inwards of the orbit of a white dwarf, which is an exceptionally small small type of star. This effect can be observed only if the white dwarf is very close to its partner, and for that to happen the partner has to be another white dwarf or a black hole. The in-spiralling comes from the fact that the pair are losing energy by emitting gravitational waves.

- What is called the gravitational redshift of spectral lines, observed in light from a white dwarf star.

- The detection of gravitational waves coming from an astronomical source.

- The so-called Cosmological Standard Model, which describes the history of the universe. It agrees with observation and it invokes General Relativity.

The last item is interesting because it led Einstein to make a proposal that he later called his biggest blunder. Here's what happened. When General Relativity first came out, the Universe was generally assumed to be static. To allow that, Einstein assigned a negative energy to the vacuum, which would cancel the positive energy coming from the content of the Universe and allow the Universe to be static. When the Universe was found to be expanding, Einstein's negative energy wasn't needed, and that's when he called the negative energy proposal his biggest blunder. He thought, retrospectively, that the idea of assigning energy to the vacuum was ridiculous.

It's a pity that Einstein didn't live to see the next development. Towards the end of the first decade of our century it was found that observation does require the vacuum to have energy. It requires that energy to be positive, so that the Universe expands faster than would otherwise be the case. So in the end, Einstein's only blunder was to choose the wrong sign for the energy of the vacuum.

Einstein took no part in the development of quantum electrodynamics (the theory of electromagnetism which takes into account quantum theory). That is perhaps surprising, because quantum electrodynamics was initiated by the English physicist Paul Dirac in 1927, when Einstein was only forty-eight years old. Neither did he pursue cosmology when Hubble announced the expansion of the universe, even though that happened when Einstein was just fifty years old. Instead, Einstein in the final thirty years of his life, tried in vain to find unified field theories, which would include both electromagnetism and gravitation and perhaps also quantum physics.

For accounts of Einstein's life and work, see [2] and [148]–[163]. For the quoted account of Einstein's relationship with music see [162]. For *Critique of Pure Reason* see [164]. For *Cosmos: A Sketch of the Physical Description of the Universe* see [165].

10.3 DAVID HILBERT

Einstein arrived at the rather complicated field equation of General Relativity, after a long process of trial and error. A contemporary called David Hilbert instead arrived at the field equation directly, using elegant mathematics. From Hilbert's viewpoint, Einstein's field equation represents the simplest possibility, that is consistent with General Relativity's description of space and time. I end with a brief account of Hilbert.

David Hilbert was born in 1862 in Königsberg in the Kingdom of Prussia (now Kalingrad in Russia), and he died in 1943 in Göttingen in Germany. He is recognized as one of the most influential and universal mathematicians of the 19th and early 20th centuries.

David Hilbert

Starting in 1886, Hilbert worked for nine years at the University of Königsberg, becoming after some years a professor. In 1895, at age thirty-three, he moved to the world's then top mathematics university, the University of Göttingen where he would spend the rest of his career. When he first arrived as a new professor, he upset the older professors by going to the local billiard hall, where he played against his juniors. He was worshiped by his students, and he would take walks with students to talk about mathematical problems.

In 1900 he put forth a list of twenty-three unsolved problems at the International Congress of Mathematicians in Paris. This is generally reckoned as the most successful and deeply considered compilation of open problems ever to be produced by an individual mathematician. Some of them were solved within a short time, but some of the others remain unsolved to this day.

Until 1912, when he was fifty, Hilbert was almost exclusively a pure mathematician. After that he instead focussed mostly on physics. In 1915 he invited Einstein to Göttingen to give a week of lectures on General Relativity. At that stage Einstein had not arrived at his field equation, and Hilbert began work on that topic. He and Einstein found the field equation independently, and published their work at almost the same time in November 1915.

In addition to his work on General Relativity, Hilbert did some fundamental work on the mathematical formulation of quantum physics. For that purpose he discovered the mathematical construction known as Hilbert space. He also tried to embed physics into what he saw as a more secure mathematical foundation, saying that 'physics is too hard for physicists'.

When Hilbert's colleague Courant wrote a book called *Methoden der mathematischen Physik* (*Methods of Mathematical Physics*), he included Hilbert's name as an author because book leaned heavily on Hilbert's ideas. The book was still the most important source for mathematical methods when I started physics research myself, and I used its English version a lot.

Around 1925 Hilbert developed pernicious anemia, a then untreatable disease whose primary symptom is exhaustion, and he did little work thereafter. Hilbert lived to see the Nazis purge many of the prominent faculty members at University of Göttingen in 1933. About a year later, Hilbert attended a banquet and was seated next to the new Minister of Education, Bernhard Rust. Rust asked whether the Mathematical Institute really suffered so much because of the departure of the Jews'. Hilbert replied, 'Suffered? It doesn't exist any longer, does it!'. When Hilbert died in 1943 at the age of eighty-one, only about ten people attended his funeral because the Nazis had more or less cleared Göttingen's mathematics faculty of people he knew.

For more on Hilbert see [166, 167].

Bibliography

[1] *Archimedes to Hawking*, Clifford A. Pickover, Oxford University Press (2008).

[2] *On giants' shoulders*, Melvyn Bragg, Hodder and Stoughton (1998).

[3] *The Sleepwalkers: A history of man's changing vision of the universe*, Arthur Koestler, Hutchingson (1959).

[4] *The works of Archimedes* T. L. Heath, ed., Mineola, N.Y. : Dover Publications (2002).

[5] *Pappus of Alexandria*, Heath, Thomas Little (1911) in Chishom, Hugh *Encyclopedia Brittanica* (11th edition), pp. 740 – 741, Cambridge University Press.

[6] *Parallel Lives*, Plutarch. *Lives: Longhorn translation*, Warne (1884).

[7] http://penelope.uchicago.edu/Thayer/E/Roman/Texts/Plutarch/home.html

[8] https://www.biography.com/people/alexander-the-great-9180468

[9] *Histories. With an English translation by W.R. Paton*, Heinemann (1922).

[10] *Ten books on architecture*, Book 9, Vitruvius, translated by Ingrid D. Howe & Thomas Noble, Cambridge University Press (1999).

[11] *Tusculan Disputations*, Cicero, translated by J. E. King. Loeb Classical Library 141. Cambridge, MA: Harvard University Press, (1927).

[12] *Loeb Classical Library, 7 volumes, Greek texts and facing English translation* translated by Charles Burton Gulick, Harvard University Press (1927-1941).

[13] `http://penelope.uchicago.edu/Thayer/e/roman/texts/` `/athenaeus.home.html`.

[14] *The Mystery of the Hanging Garden of Babylon* Stephanie Dalley, Oxford University Press (2015).

[15] `http://web.mit.edu/2.009/www/experiments/deathray` `/10_Mythbusters.html`

[16] *On the laws of the republic* Cicero, translated by Clinton W. Keyes, Loeb Classical Library 213, Harvard University Press (1928).

[17] *A History of Greek Mathematics, Volume I* Sir Thomas Heath, Courier Corporation (2012).

[18] *The World of Copernicus* Angus Armitage, EP Publishing Limited (1972).

[19] *Nicolaus Copernicus and his Epoch* Jan Adamczewski, Interpress Publishers, Warsaw (1972).

[20] *The Copernican Revolution* Thomas S. Kuhn, Vintage Books, Random House (1957).

[21] `https://kids.britannica.com/students/assembly/view/` `193178`

[22] *De Revolutionibus Orbium Celestium Libri VI*, Johannes Petreius, Nuremberg (1543).

[23] *On the Revolutions of the Heavenly Spheres* Great Books of the Western World, 16, translated by Charles Glenn Wallis, Chicago: William Benton (1952).

[24] *Copernicus: on the revolutions of the heavenly spheres*, translated by A. M. Duncan, David & Charles (Publishers) Ltd. (1976).

[25] *Ptolemy's Almagest*, translated by G. J. Toomer, Princeton University Press (1998).

[26] *Astronomia Nova ...*, Johannes Kepler, Pragae (1609)

[27] *Mysterium cosmographicum* 'The secret of the universe' Translation by A.M. Duncan / Introduction and commentary by E. J. Aiton, New York: Abaris (1981).

[28] *Johanne Kepler. Optics ...*, tranlsated by William H. Donahue, New Mexico: Green Lion Press (2000).

[29] *Dissertio cum nuncio ...*: Kepler's conversation with Galileo's *Siderial Messenger*, tranlsated by E. Rosen, Johnson Reprint Corperation (1965).

[30] *Apologia Tychonis contra Ursum*: translated by N. Jardine *The Birth of History and Philosophy and Science: Kepler's Defense of Tycho against Ursus with Essays on its Provenance and Significance* Cambridge: Cambridge University Press (1984) (with corrections 1988).

[31] *New Astronomy*, Johanne Kepler, tranlsated by W. H. Donahue, Cambridge, New York: University Press (1992).

[32] *Harmonices mundi libri V*, The Harmony of the World, Johanne Kepler, translated by E. J. Aiton, A. M. Duncan & J. V. Field, Philadelphia: American Philosophical Society (Memoirs of the American Philosophical Society) (1997).

[33] 1997. *Epitome astronomiae copernicanae* (Epitome of Copernican Astronomy: IV and V), partial translation by C. G. Wallis, , Chicago, London: Encyclopaedia Britannica (Great Books of the Western World, Volume 16], (1952).

[34] *Strena seu de nive sexangula* (The Six-Cornered Snowflake), translated by C. Hardie, Oxford: Clarendon Press, (1966).

[35] *Johannes Kepler: Life and Letters* C. Baumgardt with introduction by A. Einstein, New York: Philosophical Library (1951).

[36] *Somnium seu de astronomia lunari* The Dream, translated by E. Rosen, Madison: University of Wisconsin Press, (1967).

[37] *John Kepler*, Angus Armitage, Faber and Faber (1966).

[38] *Johannes Kepler*, M. Caspar, New York: Dover Publications (1993).

[39] *A defense of Galileo*, Thomas Campanella (1622), tranlsated by Richard J. Blackwell, University of Notre Dame Press (1994).

[40] *The martyrs of science, or the lives of Galileo, Tycho Brahe and Kepler* David Brewster, John Murray (1858).

[41] https://archive.org/stream/b24853288#page/n11/mode/2up (Martyrs online)

[42] http://galileo.rice.edu/

[43] *Discoveries and opinions of Galileo*, Stillman Drake, Doubleday (1957).

[44] *Galileo: the man, his work and misfortunes* James Brodrick, G. Chapman (1965).

[45] *Men of physics: Galileo Galilei: his life and works* Raymond J. Seeger, Pergamon Press (1966).

[46] *A long-lost letter from Galileo to Peiresc on a magnetic clock*, Bern Dibner & Stillman Drake, Norwalk, Conn. Burndy Library (1967).

[47] *Galileo, science and the church* Jerome J. Langford, University of Michigan Press (1971).

[48] *Galileo's intellectual revolution* William R. Shea, Macmillan (1972).

[49] *Galileo and Copernican astronomy: a scientific world view defined* Clive Morphet, Butterworths (1977).

[50] *Galileo at work: his scientific biography* Stillman Drake, University of Chicago Press (1978).

[51] *Galileo Studies* Alexandre Kyoré, translated by John Mepham, Humanities Press (1978).

[52] *Galileo* Stillman Drake, Oxford University Press (1980).

[53] *Galileo heretic* Pietro Redondi, translated by Raymond Rosenthal, Princeton University Press (1987).

[54] *Galileo courtier: the practice of science in the culture of absolutism* Mario Biagoli, University of Chicago Press (1993).

[55] *Galileo: decisive innovator* Michael Sharrat, Blackwell (1994).

[56] *Galileo and his sources: the heritage of the collegio romano in Galileo's science*, Willam A. Wallace, Princeton University Press (1984).

[57] *The Galileo affair: a documentary history* edited and tranlsated with an introduction and notes, by Maurice A. Finocchario, University of California Press (1989).

[58] *The Cambridge companion to Galileo* Peter Machamer, ed., Cambridge University Press (1998).

[59] *Galileo: a life* James Reston Jr., HarperCollins (2000).

[60] *Galileo: a very short introduction* Stillman Drake, Oxford University Press (2001).

[61] *Retrying Galileo* Maurice A. Finochiaro, University of California Press (2005).

[62] Letters to Élie Diodati written in 1637 and 1638, as translated in *The Private Life of Galileo: Compiled primarily from his correspondence and that of his eldest daughter, Sister Maria Celeste (1870)* by Mary Allan-Olney, Kessinger Publishing (2008).

[63] https://www.cairn.info/revue-philosophia-scientiae-2017-1-p-165.html

[64] *Science*, 16 Sep 1960: Vol 32, issue 3429, pp732.

[65] *Siderius Nuncio* Galileo, Thomas Baglioni (1610).

[66] *Siderius Nuncio* Galileo. Translated by Edward Stafford Carlos (1880), edited and corrected by Peter Barker, Byzantium Press (2004).

[67] *Kepler's conversation with Galileo's Siderial messenger* 1st complete translation with an introduction and notes, by Edward Rosen, Johnson Reprint Corporation (1965).

[68] *Discorso del flusso e reflusso del mare* Galileo, printed in *Le Operi de Galileo Galilei* Tipografia di G. Barbèra (1890). Translated in *The Galileo Affair: A Documentary History* Maurice Finocchiaro, University of California Press (1989).

[69] *Istoria e Dimostrazioni intorno alle Macchie Solari* Galileo, Accademia de Lincei (1613). Translated in *On Sunspots* Eileen Reeves & Albert Van Heiden, University of Chicago Press (2010).

[70] *Dialogue Concerning the Two Chief World Systems–Ptolomaic & Copernican* Galileo (1632), translated by Stillman Drake, University of California Press (1962).

[71] *Dialogues Concerning Two New Sciences* Galileo (1638). Translated by Alfonso de Salvio & Henry Crew, The MacMillan Company (1924).

[72] http://oll.libertyfund.org/titles/galilei-dialogues-concerning-two-new-sciences

[73] http://galileo.rice.edu/sci/instruments/thermometer.html

[74] https://longitudeprize.org/about-us/history

[75] https://www.space.com/14736-sunspots-sun-spots-explained.html

[76] http://physics.kenyon.edu/EarlyApparatus/Fluids/Westphal_Balance/Westphal_Balance.html

[77] http://www.csun.edu/~hbund408/math%20history/decimal.html

[78] *Newton: the man* Richard De Villamil, G. D. Knox (1931).

[79] *Isaac Newton – a biography, etc with a portrait* (1934) by Louis Trenchard More (1870–1944), citing unpublished papers by John Conduitt (1688–1737).

[80] *Newton demands the muse* Marjorie Hope Nicolson, Princeton University Press (1946).

[81] *Isaac Newton: historian* Frank Edward Manuel, Cambridge University Press (1963).

[82] *Isaac Newton* J. D. North, Oxford University Press (1967).

[83] *A portrait of Isaac Newton* Frank E. Manuel, Harvard University Press (1968).

[84] *Never at rest: a biography of Isaac Newton* Richard S. Westfall, Cambridge University Press (1983).

[85] *Let Newton be!* Ed. John Fauvel. Oxford University Press (1988).

[86] *The life of Isaac Newton* Richard S. Westfall, Cambridge University Press (1993).

[87] *The religion of Isaac Newton* Frand E. Manuel, Clarendon Press (1974).

[88] *Popular Astronomy: a general of the heavens*, Camille Flammarion, translated by J. Ellard Gore, Chatto & Windus (1894).

[89] *A short account of the history of mathematics*, W. W. Rouse Ball, (reprint of original of 1908) Dover Publications (1960).

[90] https://cudl.lib.cam.ac.uk/view/MS-ADD-03996/8

[91] *Mathematical Papers of Isaac Newton. Volume I 1664–66* Ed. D. T. Whiteside, assisted by M. A. Hoskin, Cambridge University Press (1967).

[92] https://www.gutenberg.org/files/33504/33504-h /33504-h.htm

[93] *The optical papers of Isaac Newton* Ed. Alan E. Shapiro, Cambridge University Press (1984).

[94] *Philosophiæ Naturalis Principia Mathematica*, Benjamin Motte (1687).

[95] https://royalsociety.org/collections/principia-mathematica/

[96] https://cudl.lib.cam.ac.uk/view/PR-ADV-B-00039-00001/1

[97] *The Principia: The Authoratative Translation and Guide*, translated by I. Bernard Cohen, University of California Press (2016).

[98] *Introduction to Newton's Principia* I. Bernard Cohen, Harvard University Press (1971).

[99] *The preliminary manuscripts for Isaac Newton's 1687 Principia, 1684-1685: facsmile of the original autographs, now in Cambridge University Library* Derk T. Whiteside, Cambrdige University Press (1989).

[100] *Newton on matter and activity* Erman McMullin, University of Notre Dame Press (1978).

[101] *Isaac Newton and gravity* P. M. Rattansi, Wayland (1980).

[102] *The mathematical papers of Isaac Newton* Ed. D. T. Whiteside with the assistance in publication of M. A. Hoskin, Cambridge University Press (1967–1981).

[103] Newton. Philosophical Writings Andrew Janisk ed. Cambridge University Press (2004).

[104] *The library of Isaac Newton* John Harrison, Cambridge University Press (1978).

[105] http://www.newtonproject.ox.ac.uk/texts/correspondence/optical

[106] *The correspondence of Isaac Newton, vol 1* H. W.Turnbull ed. Cambridge University Press (1959). *The correspondence of Isaac Newton, vol 2* H. W. Turnbull ed. Cambridge University Press (1960). *The correspondince of Isaac Newton, vol 3* H. W. Turnbull ed. Cambridge University Press (1961).

[107] http://www.newtonproject.ox.ac.uk/texts

/newtons-works/all

[108] http://www.newtonproject.ox.ac.uk/view/texts/
normalized/THEM00033

[109] *Memoirs of Sir Isaac Newton's life* William Stuckeley & A. I.
Ellis, Alfred Hastings White (1752).

[110] *Memoirs of Sir Isaac Newton's life* William Stuckeley & A. I.
Ellis, Taylor & Francis (1936).

[111] http://www.bbc.co.uk/history/british/timeline/
civilwars_timeline_noflash.shtml

[112] https://www.uu.edu/centers/science/voice/article
.cfm?ID=29

[113] E. Spinak and A. Packer *350 years of scientific publication: from
the "Journal des Sçavans" and Philosophical Transactions to Sci-
ELO [online]*. SciELO in Perspective, 2015 [viewed 20 February
2018].

[114] http://blog.scielo.org/en/2015/03/05/
350-years-of-scientific-publication-from
-the-journal-des-scavans-and-philosophical
-transactions-to-scielo/

[115] http://rspl.royalsocietypublishing.org/

[116] https://www.999inks.co.uk/british-newspaper-history
.html

[117] https://www1.umassd.edu/ir/resources/laboreducation
/literacy.pdf

[118] https://www.cdc.gov/plague/history/index.html

[119] *Oersted and the discovery of electromagnetism* Bern Dibner,
Blaisdell Publishing Company (1962).

[120] https://www.famousscientists.org/hans

-christian-oersted

[121] https://www.aps.org/publications/apsnews/200807/
physicshistory.cfm

[122] https://www.thefamouspeople.com/profiles/hans
-christian-oersted-6510.php

[123] *Michael Faraday* H. G. Andrew, published for the Nuffield Foundation by Longmans/Penguin Books, Longmans Green and Co. Ltd (1966).

[124] *Faraday Redescovered* Ed. David Gooding & Frank A.J.L. James, The Macmillan Press Ltd. (1985).

[125] *The Life and Letters of Faraday* Dr. Bence Jones, Longmans, Green and Co. (1870).

[126] *Michael Faraday: a very short introduction* Frank A.J.L. James, Oxford University Press (2010).

[127] *Makers of electricity* Michael O'Reilly and James Walsh, New York Fordham University Press (1909).

[128] *André-Marie Ampère* James R. Hofmann, Cambridge University Press (1995).

[129] *Le Grand Ampère* Louis de Launay, Librarie Académique Perrin (1925).

[130] *Corpus poetarum latinorum*, William Sydney Walker, Londini, H. G. Bohn (1849).

[131] *Émile ou de l'éducation* Jean-Jaques Rousseau (1762).

[132] *Émile ou de l'éducation* Jean-Jaques Rousseau, Gallimard (1995).

[133] *Emile or on Education* translation by Allan Blook, Basic Books (1979).

[134] http://oll.libertyfund.org/titles/rousseau-emile
-or-education

[135] *Méchanique Analitique* Joseph-Louis Lagrange (1788)

https://www.irphe.fr/~clanet/otherpaperfile/
articles/Lagrange/N0029071_PDF_1_530.pdf

[136] *Analytical Mechanics* Joseph-Louis Lagrange, translated by Auguste Boissonnade & Victor N. Vagliente, Springer (1997).

[137] *Oeuvres de Lagrange* Joseph-Louis Lagrange, Gauthier-Villars (1867).

[138] *Encyclopédie* ed. Denis Diderot & Jean le Rond d'Alembert, Antoine-Claude Briasson (1751–66).

[139] *Popular Astronomy: A General Description of the Heavens* Camille Flammarion, Chatto and Windus (1894).

[140] *Eulogy of Descartes* translated from the French by M. Thomas, Cheltenham (1826).

[141] *A dynamical theory of the electromagnetic field*, James Clerk Maxwell, London, the Royal Society (1865).

[142] *The scientific papers of James Clerk Maxwell* Ed. W. D. Niven, Dover Publications (1965).

[143] *The life of James Clerk Maxwell* Lewis Campbell & William Garnett, with a new preface and appendix with letter, by Robert H. Kargon, Johnson Reprint Corporation (1969).

[144] *James Clerk Maxwell: a biography* Ivan Tolstoy, Canongate (1981).

[145] *The demon in the aether: the story of James Clerk Maxwell* Martin Goldman, A. Hilger (1983).

[146] *A natural philosophy of James Clerk Maxwell* P. M. Harman, Cambridge University Press (1998).

[147] *James Clerk Maxwell: perspectives on his life and work* Ed. Raymond Flood, Mark McCartney & Andrew Whitaker, Oxford University Press (2014).

[148] *Albert Einstein, philosopher-scientist* Paul Arthur Schilipp, Harper & Row (1949–1951).

[149] *Einstein on peace* Ed. Otto Nathan & Heinz Norden. Preface by Bertrand Russell, Shocken Books (1960).

[150] *Einstein, the man and his achievements: a series of broadcast talks* under the general editorship of G. C. Whitrow, British Broadcasting Coroporation (1967).

[151] *Conversations with Einstein* Alexander Moszkowski, translated by Henry L. Brose; introduction by Henry LeRoy Finch; foreword by C. P. Snow, Sidgwick & Jackson (1972).

[152] *Einstein; the life and times* Ronald W. Clark, Hodder & Stoughton (1973).

[153] *Albert Einstein: creator and rebel* Banesh Hoffmann; with the collaboration of Helen Dukas, Hart–Davis, MacGibbon (1973).

[154] *Einstein's universe* Nigel Calder, British Broadcasting Corporation (1979).

[155] *Einstein for beginners* Joe Schwartz & Michael McGuinness, Pantheon Books (1979).

[156] *Einstein, the human side: new glimpses from his archives* selected and edited by Helen Dukas & Banesh Hoffmann, Princeton University Press (1979).

[157] *"Subtle is the Lord": the science and the life of Albert Einstein* Abraham Pais, Oxford University Press (1982).

[158] *The young Einstein: the advent of relativity* Lewis Pyenson, Adam Hilger (1985).

[159] *Einstein* Jeremy Bernstein, Fontana (1991).

[160] *Albert Einstein and the frontiers of physics* Jeremy Bernstein, Oxford University Press (1996).

[161] *Einstein and religion: physics and theology* Max Jammer, Princeton University Press (1999).

[162] Article by Leon Botstein in *Einstein for the 21st Century: His Legacy in Science, Art, and Modern Culture*, Ed. Peter L. Galison, Gerald Holton & Silvan S. Schweber, Princeton University Press (2008).

[163] *Einstein: a biography* Jürgen Neffe, Farrar, Straus & Giroux (2007), translated by Shelley Frish, John Hopkins University Press (2009).

[164] *Critique of Pure Reason* Emmanuel Kant,
https://www.gutenberg.org/files/4280/4280-h/4280-h.htm

[165] *A Sketch of the Physical Description of the Universe, Vol. 1,* Alexander von Humbolt
http://www.gutenberg.org/ebooks/14565

[166] https://www.britannica.com/biography/David-Hilbert

[167] http://www.storyofmathematics.com/20th_hilbert.html

Index

Printed in the United States
by Baker & Taylor Publisher Services